Signals and Communication Technology

For other titles published in this series, go to
www.springer.com/series/4748

Sofie Pollin · Michael Timmers ·
Liesbet Van der Perre

Software Defined Radios

From Smart(er) to Cognitive

Springer

Sofie Pollin
SSET/wireless
IMEC
Kapeldreef 75
Leuven 3001
Belgium
pollins@imec.be

Dr. Liesbet Van der Perre
IMEC VZW
Kapeldreef 75
Leuven 3001
Belgium
vdperre@imec.be

Michael Timmers
Bell Labs
Alcatel-Lucent
Copernicuslaan 50
Antwerpen 2018
Belgium
michael.timmers@alcatel-lucent.com

ISSN 1860-4862
ISBN 978-94-007-3597-2 ISBN 978-94-007-1278-2 (eBook)
DOI 10.1007/978-94-007-1278-2
Springer Dordrecht Heidelberg London New York

Cover design: VTeX UAB, Lithuania

Printed on acid-free paper

Springer is part of Springer Science+Business Media (www.springer.com)

Preface

The perfect engine does not win the race. Similarly, Software Defined Radio and Opportunistic Spectrum Access open up great opportunities to realize ubiquitous wireless services. However, only smart(er) operation of these engines will result in truly benefiting from them. In this book we aim to introduce and apply a practical design approach towards smart(er) and cognitive radios. We are grateful that Springer is willing to publish this book on smart(er) and cognitive radios, whereby we don't want to claim other radios (books) were dumb. Dear reader, we hope you may find some ideas of interest to your work or study, or maybe a case that could help improve your products. We want to acknowledge our colleagues at IMEC, Bell Labs and their networks for their great scientific contribution, and the enlightening discussions, both technically and way beyond. This book's creation faced fierce competition from our busy professional occupation, and 'rush-hour' in our personal life. Two babies left the 'design phase' to go in 'real-life operation', and even run-time, while three other children showed to be running ever faster, and evolving to 'advanced cognitive' behavior. We thank Liselore, Seppe, Stien, Nore and Sara for the inspiration they bring in our lives.

Leuven

Sofie Pollin
Michael Timmers
Liesbet Van der Perre

Contents

List of Acronyms

TRM	Terminal Reconfiguration Manager
OSM	Operator Spectrum Manager
TRC	Terminal Reconfiguration Controller
RAN	Radio Access Network
TMC	Terminal Measurement Collector
RRC	RAN Reconfiguration Controller
RMC	RAN Measurement Collector
NRM	Network Reconfiguration Manager

List of Figures

List of Tables

Chapter 1
Serving Many Mobile Users in Various Scenarios: Radios to Go Smart(er) and Cognitive

1.1 Towards Cognitive Radio

Anything, anytime, anywhere! The holy grail of wireless communication is coming closer. However, significant hurdles still need to be taken. Indeed, *Anytime* would require the systems to have an infinite amount of energy resources, so as to guarantee an endless operation. Further, *Anything* implicitly means that the Quality of Service (QoS) guarantees are as tough as they are diverse. Finally, *Anywhere* would require the systems to be able to operate effectively in a broad range of heterogeneous environments.

The main difficulty behind the AAA paradigm lies in the scarcity of the used resources. Thus, it is critical to share this resource efficiently and effectively. In this book, we focus on wireless spectrum and energy as scarce resources and which are, hence, becoming expensive. Moreover, the increased flexibility in radio hardware and wireless standards is both opening new opportunities and posing new challenges. As a result, radios need to evolve to smart and cognitive.

In this introductory chapter, we embark by highlighting how HardWare (HW) implementation and policy guidelines are becoming increasingly flexible. We then present the Cognitive Radio (CR) framework that implements the necessary control functionality to harness this flexibility. The need for smart and cognitive radios is explained, and the way forward as proposed in this book is introduced.

1.2 Increasing the Hardware Flexibility

The current and emerging wireless telecommunication landscape is pushing for more hardware flexibility. Software Defined Radios (SDRs) give an answer to this need.

S. Pollin et al., *Software Defined Radios*, Signals and Communication Technology,
DOI 10.1007/978-94-007-1278-2_1, © Springer Science+Business Media B.V. 2011

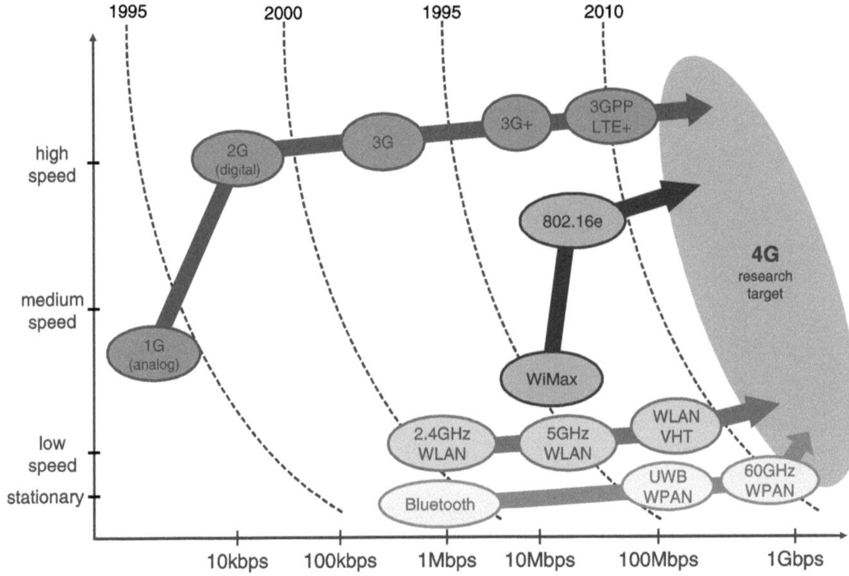

Fig. 1.1 Standards continue to accommodate higher mobility and achieve higher throughput. To accommodate the wishes of the wireless user, a divergence, rather than a convergence, of wireless standards, can be seen today

1.2.1 Wireless Landscape Giving Challenges and Opportunities

Wireless standards continue to diverge, driven by heterogeneous user and application requirements. Increased flexibility in radios may offer the long desired seamless connectivity. This increased flexibility is not only a functional wish, due to the increasing cost of ASIC designs it also becomes an economic necessity [1].

1.2.1.1 Heterogeneity Desires Flexibility

Wireless communications are routinely used today for a large variety of applications, including voice, data transfer, Internet access, audio and video streaming and social networking. For one specific service, several systems have been standardized and some have become the preferred option in several regions of the world.

Pushed by the insatiable demand for bandwidth and pulled by the steady improvement of semiconductor technology, the performance delivered by wireless standards is destined to improve, seemingly without bounds (see Fig. 1.1). To accommodate the diverse wishes of the wireless user, we are witnessing a divergence, rather than convergence, of wireless standards. This is especially true in terms of various modes of use.

From a user perspective, it is, hence, attractive to have a single handheld device that can support a large variety of wireless standards. This ensures interoperability and gives the user access to a wide variety of applications. As a result, mobile handsets have started supporting multiple modes over the past years. Initially, this has been achieved by integrating multiple radios into one handset. However, when the number of radios increases, the cost, size, and weight of the terminal are also affected. Hence, a single radio that comprises all necessary functionality is highly sought-after [1, 2]. Users want to enjoy a multitude of services on one terminal. Predictions claim that handheld devices need to support at least six different radios at the short term. Hence, also from a functionality point of view, flexibility is becoming essential.

1.2.1.2 Enabling Seamless Connectivity

Ubiquitous and seamless connectivity can be achieved in a heterogeneous network environment, under the condition that both terminals and network feature the necessary reconfiguration capabilities to support horizontal (between access points of one technology) and vertical (between access points supporting different technologies) roaming. Recently, the need for such reconfiguration support is receiving attention in specific standardization initiatives (see Chap. 2).

1.2.1.3 Scaling Technology Imposes Reconfigurability

Gordon E. Moore predicted 45 years ago, that the number of transistors on a chip would double about every 2 years. So far, history has proved him correct [3]. Scaling has brought enormous processing capabilities packed on small areas, opening opportunities to implement flexible platforms at low cost and low power.

However, for newer technologies, the Non-Recurring Engineering (NRE) costs related to System-on-Chip (SoC) design are rising exponentially. On top of this, design cost has increased dramatically and is expected to continue to do so. Not only is the design complexity increasing, CMOS scaling has arrived at the point where parasitic problems are becoming dominant. As these effects cannot be resolved at the transistor side, new designs will need to take these problems into account.

The question whether scaling is still viable is sounding louder each day, especially for custom ASIC chips. While higher production volumes can still compensate the NRE costs, cost trade-offs show, already today, an advantage in using reconfigurable radios to single-mode devices for smaller markets. For these instances, the extra area penalty is not significant compared to the cost cutting in NRE.

The extra area penalty can be furthermore easily compensated for multi-mode terminals. In those terminals the possibility exists to reuse silicon and thus significantly reduce overall area, which makes flexible platforms even more attractive. Also other cost factors, like assembly cost, form factor and time-to-market, direct towards chip reuse and, hence, the HW platform paradigm [4, 5].

Table 1.1 SDR Forum's 5-tier concept [7]

Tier	Name
0	Hardware Radio
1	Software Controlled Radio
2	Software Defined Radio
3	Ideal Software Radio
4	Ultimate Software Radio

1.2.2 The Software-Defined Radio Solution

As mentioned above, only flexible radios enable the AAA paradigm, as their flexibility allows them to operate in any environment and under any user request. This observation gave birth to the concept of Software Radio (SR), an extremely flexible radio. Today, the SR-concept is well established, but the ultimate SR (uSR) is still not in reach. According to the SDRForum such an uSR accepts fully programmable traffic and control information, supports operation over a broad range of frequencies and can switch from one air-interface to another in milliseconds [6]. Though the uSR might not yet be in reach, the research community has already made significant advances to increase the hardware flexibility.

In Table 1.1, the initial 5-tier concept of the SDRForum is presented [7]. The early radios were baseline radios with fixed functionality, called Hardware Radios (HRs). Today, it may be argued that virtually all modern wireless communications equipment can be classified as Software Controlled Radios. These radios implement the signal path using application-specific hardware, i.e., the signal path is essentially fixed. A software interface may allow certain parameters to be changed in software. For most applications, the state-of-the-art flexible radios are Software Defined Radios (SDRs) (i.e. Tier 2). These radios allow the signal path to be reconfigured in software without requiring any hardware modifications. As the uSR is considered to be the *blue-sky* vision of SDR, the next target is the ideal Software Radio (iSR). Compared to a standard SDR, an iSR implements much more of the signal path in the digital domain. Ultimately, programmability would extend to the entire system with Analog/Digital Conversion (ADC) taking place at the antenna.

1.3 Increasing the Policy Flexibility

Regulatory bodies are being pushed to define spectrum access policies more flexible. The Dynamic Spectrum Access (DSA) concept, the ultimate goal of flexible spectrum access, ideally could realize the optimal usage of the spectrum.

1.3.1 Spectrum: A Scarce Resource

Like water, air or oil, the wireless spectrum is a shared resource. This, however, does not imply that it should be free! Wireless spectrum needs to be shared with many applications, and, hence, it has become expensive. To deliver the required expensive service, regulatory bodies are using a fixed frequency allocation scheme, out of fear of harmful interference, which jeopardizes the quality of the delivered service. They allocate spectrum blocks for multiple years and over large areas. The only exception to this allocation scheme, is the ISM band, where heterogeneous devices can coexist using high-level Listen-Before-Talk (LBT) etiquettes.

Due to this fixed frequency allocation scheme, along with the accelerated deployment of broadband communication systems, spectrum is, hence, becoming a major bottleneck. New applications require more and more spectrum, but no useful spectrum is apparently left to be allocated (see upper part of Fig. 1.2).

However, experiments show that up to 85% of the spectrum remains unused at a given time and location, indicating that a more flexible allocation strategy could solve the spectrum scarcity problem [8]. We have confirmed this through measurements using the IMEC Scaldio [9] (see bottom part of Fig. 1.2).

This inefficient use of spectrum and the success of the ISM bands has coerced regulatory bodies into defining a roadmap towards more local, organic and dynamic spectrum sharing policies [10–12].

1.3.2 The Opportunistic Spectrum Access Solution

As mentioned above, a need for policies to share spectrum dynamically exists. A prime example of such a flexible policy is Opportunistic Spectrum Access (OSA), where users can actively search for unused spectrum in licensed bands and communicate using these *white holes*. OSA is supported by regulatory bodies, such as the Federal Communications Commission (FCC) [10] and the European Commission (EC) [12]. The concept is also often ambiguously referred to as Cognitive Radio (CR). We discuss the ambiguity of this terminology in Sect. 1.4.

The ultimate goal of the roadmap for more local and organic spectrum policies, is called Dynamic Spectrum Access (DSA). In DSA networks licenses and priorities are not fixed at Design Time (DT). DSA networks should allow terminals and technologies to negotiate the use of wireless spectrum locally for a time window of hours, minutes or even seconds. Like the SR, which presents the ultimate flexible radio, DSA is not yet in sight. However, OSA is a first disruptive step towards this ultimate goal.

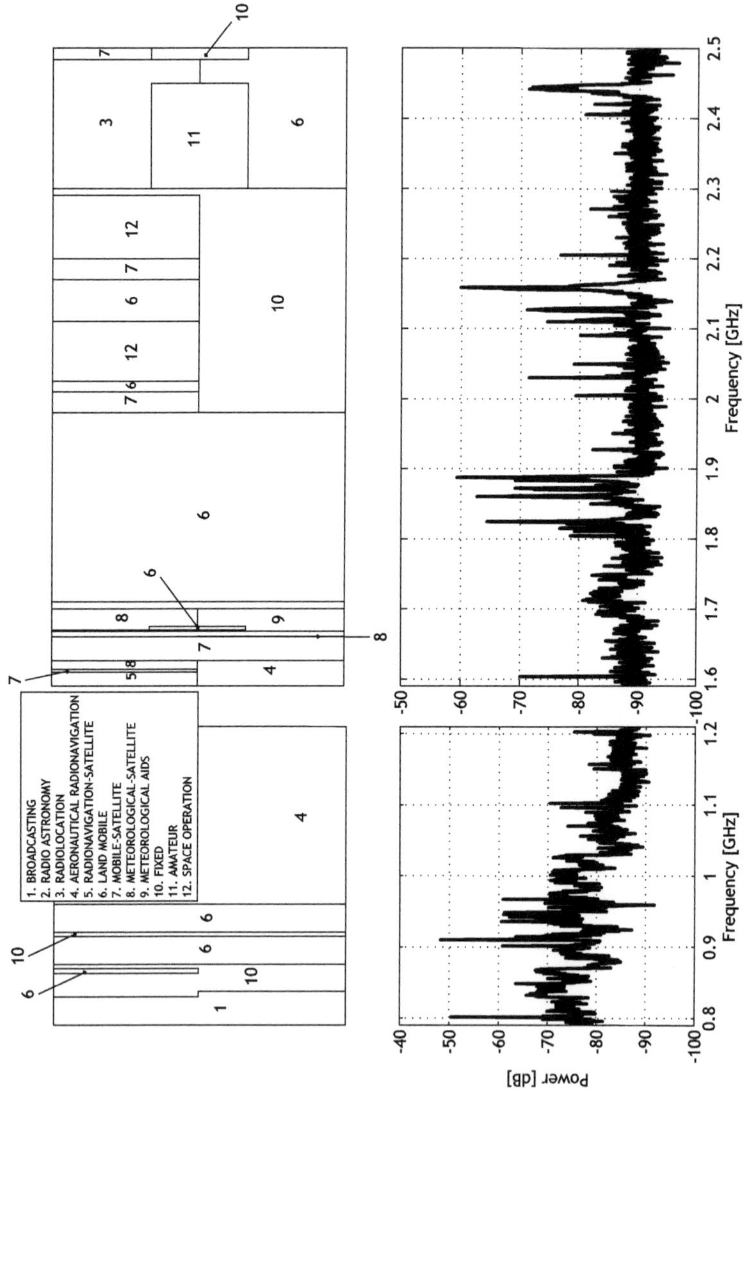

Fig. 1.2 In the *upper part*, the current Belgian spectrum plan is shown [13]. Spectrum appears to be very scarce as no bands are apparently left free to allocate. However, the *lower part* shows measurements of the same frequency range taken by the IMEC Scaldio chip on at 13h15, 6th of July 2009. When we take a snapshot at a certain time and location, a lot of this licensed spectrum is not being used. Indeed, only the popular standards seem to be semi-densely used. However, measurements by TU Berlin have shown that even for the extremely popular GSM standard OSA remains viable [14]

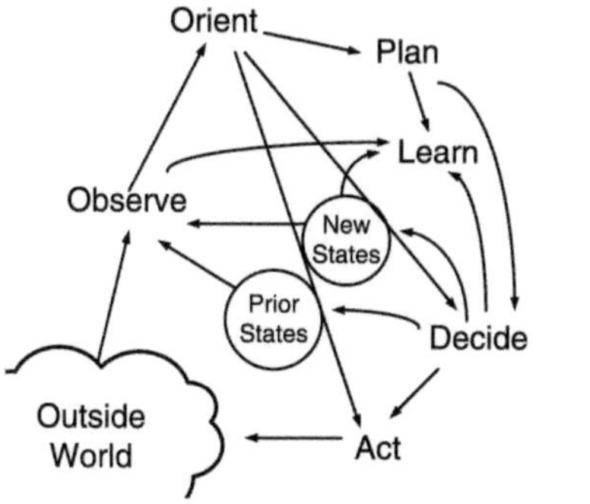

(a) According to Mitola, the CR is a natural extension of the SDR. Centrally, in his cognition cycle is an intelligent control mechanisms, that is able to learn from the environment [15].

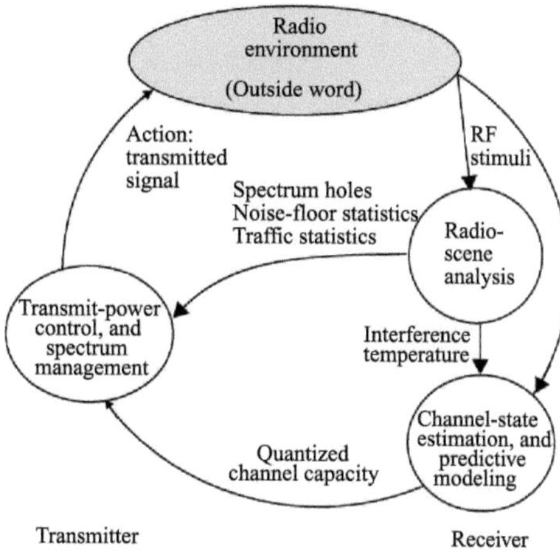

(b) Though Haykin largely agreed with Mitola's definition, he instantiated his cognition cycle for OSA. In this cognition cycle only 2 key CR elements remain: environment awareness and adaptivity [17].

Fig. 1.3 A comparison between the cognition cycles of Mitola and Haykin

1.4 Cognitive Radio: Exploiting Flexibility with Intelligent Control

The Cognitive Radio (CR) concept brings an answer to exploit the increased flexibility with smart(er) control solutions.

The CR was first described by Mitola in [15, 16] as a decision making layer in which "*wireless personal digital assistants and the related networks were sufficiently computationally intelligent about radio resources, and related computer-to-computer communications, to detect user needs as a function of use context, and to provide radio resources and wireless services most appropriate to those needs*". Through the eyes of Mitola CR was a natural extension of the SDR. His CR cycle can be seen in Fig. 1.3(a).

Six years after Mitola's first CR article, Simon Haykin recapitulated the CR idea as an enabler of brain-empowered communication [17]. He identified six key parts of CR: awareness, intelligence, learning, adaptivity, reliability and efficiency. Thus, his definition broadly complies with Mitola's viewpoint. In his seminal paper, Haykin instantiated CR for OSA as can be seen in Fig. 1.3(b). In this simplified cognition cycle only 2 key elements of his CR are instantiated: environment awareness (radio-scene analysis and channel estimation) and adaptivity (transmit power control and dynamic spectrum management).

Together with the initial focus of CR research on enabling environmental awareness through spectrum sensing, this led to people using CR as an equivalent of OR.

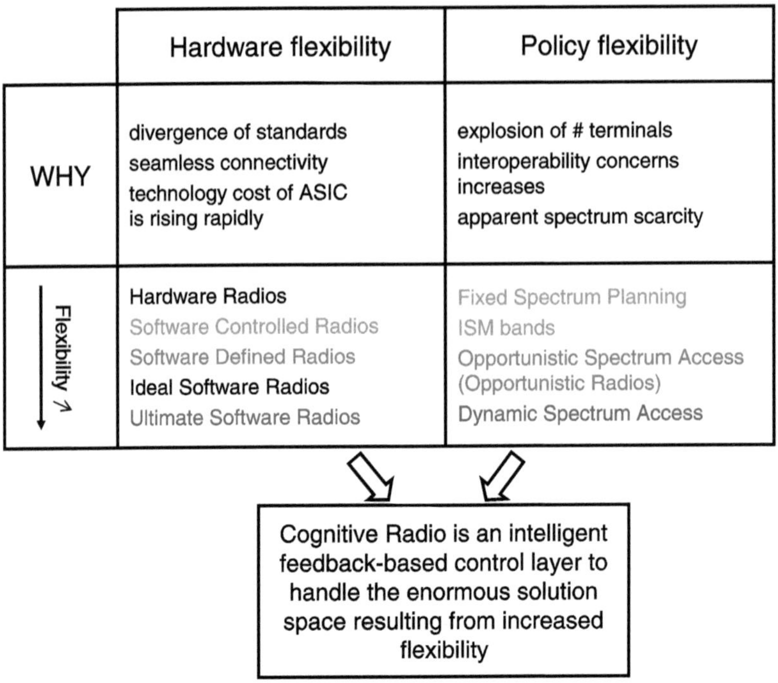

Fig. 1.4 The Cognitive Radio is an adaptive feedback-based layer to control the increasing flexibility, of which prime examples are the SDR (at the hardware side) and OR (at the spectrum side)

For instance, IEEE 802.22 was marketed as the first CR standard, while it actually carries few of the key elements present of the CR as defined by Mitola or Haykin.

Mitola and Haykin actually saw the CR as an answer to the following questions:

- "What can we do with the increasing HW flexibility provided by the SDR?" (Mitola, 1999).
- "What can we do with the increasing policy flexibility provided by regulators?" (Haykin, 2004).

The common denominator of these two visions is to define a CR as an *intelligent control framework to handle the increased flexibility*.

To avoid ambiguity, throughout this book, the spectrum-oriented view on CR (OSA) is denoted as Opportunistic Radio (OR). CR denotes an adaptive feedback-based layer to control the increasing flexibility, of which prime examples are the SDR (at the hardware side) and OR (at the spectrum side). This can be seen in Fig. 1.4. An adaptive radio that exploits flexibility to improve on QoS or energy cost, but does not learn the optimal configuration based on feedback, will be denoted as a smart radio.

1.5 The Need for a New Approach

This book proposes a framework that is able to exploit the flexibility offered by the disruptive radio hardware and spectrum access paradigms. The main problem considered is the sharing of wireless spectrum. To intelligently share this spectrum, terminals need to interact with their actual environment to learn its behavior and adapt accordingly. As mentioned in Sect. 1.4, a CR system heavily relies on the existence of a feedback channel from the environment. Hence, the proposed framework is fully compatible with the CR paradigm. In the Mitola vision it is theoretically proposed to learn everything in real time. For practical applications, we introduce a framework that is able to extract useful information upfront, when designing the system. Consequently, the performance in operation can be improved greatly. In [18, 19] it is demonstrated that a hybrid Design Time (DT)–Run Time (RT) framework allows for very efficient operations. In such DT-RT framework, procedures are developed at DT. At RT, these procedures can be efficiently executed by monitoring the environment and selecting the appropriate procedure. However, due to the increasing flexibility of wireless terminals, it is becoming harder to predict all possible situations. Furthermore, due to the increasing environment dynamics, it is also becoming harder to model these situations. Our framework should hence be flexible enough to allow calibration and learning.

We propose an extended DT-RT framework and maintain its efficient operation. In line with the CR paradigm, the framework is made flexible enough so the terminal can change its behavior based on interaction with the environment. Feedback-based learning is enabled in the framework, fully in line with the CR definitions introduced earlier.

1.6 Radios to Go Smarter and Cognitive

In this book, we aim to introduce how Software Defined Radios can be conceived to become smart(er) and cognitive. Indeed the opportunities brought about by the improved flexibility and degrees of freedom, bring along an increased control complexity. Exploiting this complexity is a challenge, which holds the opportunity to use scarce resources spectrum and energy in an optimal way. The optimization itself should work in practical cases and come at minimal overhead.We thereto sketch the standardization scene, introduce methodological concepts, and discuss the application to relevant case studies:

- **Chapter 2** discusses some of the major wireless standards to illustrate how increased flexibility and control is becoming abundant, even within a single wireless technology. Next to flexibility within standards, focus is on coexistence between standards and finally cooperation across standards. The latter often requires a flexibility across standards and increased intelligence. This drives the need for SDR and finally CR.

- **Chapter 3** introduces a general control strategy that is able to adapt flexibly to the environment without heavily impacting the run-time complexity. The method proposed consists of a preparation ('design time') and an operation phase ('run time process'). The degree of intelligence and adaptation enabled in the run time process typically determines how *smart* the radio is. It will be shown that the run time involves four main tasks: monitoring or observing, determining the current scenario, acting on the scenario following a procedure and finally learning or calibrating the procedure. In the subsequent chapters, case studies will be presented that involve each of the four tasks, but however, emphasis will be on one of the tasks.
- **Chapter 4** discusses a scenario where licensed users share their resources with opportunistic radios. In this case, the OR needs to sufficiently monitor the actual situation to avoid interference to the licensed technology. In Chap. 4 the flow will hence be applied to a OR scenario with special focus on the monitoring.
- **Chapter 5** discusses a coexistence situation. Specifically, it considers the scenario where an IEEE 802.15.4 networks coexists with an IEEE 802.11 network in the ISM band by adapting its channel. In this case, monitoring is shown not to be sufficient and determining the actual situation is based on both monitoring and learning based on feedback from the environment. It is shown how nevertheless, is possible to instantiate the flow to design a radio that is capable to adapt based on an identification of the actual scenario.
- **Chapter 6** gives an example of how the flow can be instantiated on a well-defined procedure that does not necessary involve a CR or OR. The context will be an IEEE 802.11 network in which multiple users coexist and have to share the medium, while meeting performance constraints and minimizing energy cost. Even for such an IEEE 802.11 network, a flexible radio within a network of flexible radios can be managed smartly to improve the overall QoS or energy efficiency of the network.
- **Chapter 7** then gives an example of the most important task of a cognitive radio: learning and calibrating how it behaves in the environment. How to achieve this efficiently is illustrated in the context of IEEE 802.11 networks that aim at minimizing the co-channel interference in a distributed way by adapting the output power, sensing threshold and transmission parameters.
- **Chapter 8** finally closes the book with major conclusions and a glance at the future.

Chapter 2
Emerging Standards for Smart Radios: Enabling Tomorrow's Operation

2.1 Standards in Evolution

In this chapter we detail the standardization roadmap towards an increase in hardware and policy flexibility, described in Chap. 1. This is shown in Fig. 2.1.

First, we discuss the increase in hardware flexibility. While standards typically do not discuss implementation-related aspects, we describe how parameters have become more and more dynamic. This trend is clearly driven by the innovation in the SDR, which is a reconfigurable platform solution that supports a range of air interfaces. Without flexible implementation methodologies, this level of flexibility could not have been introduced by standardization bodies. As shown in Table 2.1, we discuss standards for both cellular and distributed technologies.

Afterwards, we discuss the increase in policy or spectrum access flexibility. As shown in Chap. 1, the ISM bands, where unlicensed devices coexist using only high-level spectrum etiquettes, have been a clear driver towards more dynamic spectrum access. In this chapter, we discuss two ways to counter the QoS problem with dynamic spectrum access. First, we discuss how the spectrum access is taken a step back, by introducing more control mechanisms between the different standards in the ISM bands. Specifically, we discuss the IEEE 802.11h standard, where algorithms are developed to avoid interference from unlicensed ISM bands towards satellite and radar systems and the IEEE 802.19 standard, which takes care of cross-technology coexistence. The second way to counter the QoS problem of dynamic spectrum access is to introduce a more hierarchical scheme, known as opportunistic spectrum access. In opportunistic spectrum access, unlicensed devices can use licensed spectrum as long as they do not interfere with licensed users.

In Chap. 1, we have discussed the CR as an intelligent control layer to handle the increasing flexibility, introduced by SDR and OSA. In this chapter, we show two standardization efforts that are in line with this view. We first present the IEEE 802.21 Media-Independent Handover, which dictates how the a reconfigurable radio can decide to switch between different standards. Afterwards we discuss the IEEE DYSPAN Standards Committee (formerly SCC41), where techniques are standardized in line with the CR paradigm to allow for a more advanced adaptation and monitoring for Cognitive Radios.

S. Pollin et al., *Software Defined Radios*, Signals and Communication Technology, DOI 10.1007/978-94-007-1278-2_2,

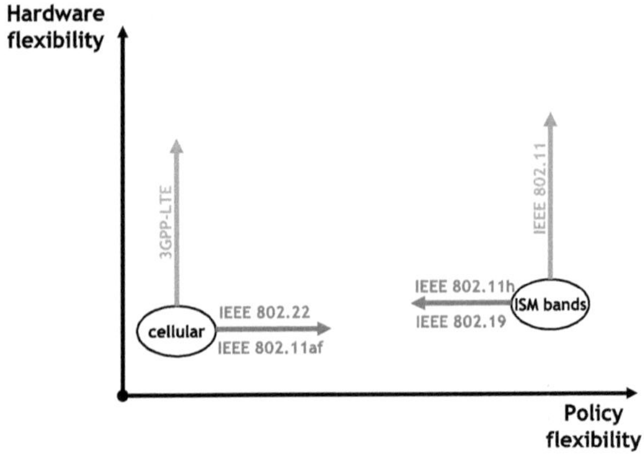

Fig. 2.1 Standards in evolution: hardware flexibility, policy flexibility towards true cognitive control

2.2 Hardware Flexibility

Some of the major wireless standards are discussed to illustrate how increased flexibility and control is becoming abundant, even within a single wireless technology. This demonstrates the diversity of the wireless scene, where many different air interfaces and applications coexist. It is shown in Sect. 2.2 that within each standard, the number of communication modes is increasing with each generation. While standardization bodies do not discuss implementation, the number of modes is increasing by such an amount that dedicated ASIC implementations become unrealistic. First, the IEEE 802.11 generations are discussed focusing on the flexibility and control options within the standard. Afterwards, we discuss 3GPP-LTE as the most prominent candidate for next generation cellular technology. Again, for this class of standards, a trend exists towards more flexibility and distributed control.

Table 2.1 Standards in evolution: hardware flexibility, policy flexibility towards true cognitive control

Radio	Scenarios	Standard
Hardware Flexibility	WLAN	IEEE 802.11
	Cellular/Licensed	3GPP-LTE
Spectrum Access Flexibility	Coexistence in Unlicensed Bands	IEEE 802.11h and 802.19
	Opportunistic Radio	IEEE 802.22, 802.11af
Cognitive Radio Control	Medium Independent Handover	IEEE 802.21 MIH
	Terminal and Network Control	IEEE DYSPAN, ETSI RSS

2.2.1 *IEEE 802.11: A Flexible Radio Becomes Smarter*

Wireless Local Area Networking (WLAN) has always been the most important technology for wireless data solutions, although competition from the next generation cellular technologies is increasing. The main standard for those WLAN networks has been the IEEE 802.11 standard since 1999.

IEEE 802.11 specifies a connectionless MAC on top of multiple PHY layers in the unlicensed bands at 2.4 GHz and 5 GHz and recently also the 60 GHz band [20]. In this chapter, we first give a short overview of the IEEE 802.11 standards in the unlicensed bands below 6 GHz. Afterwards, the flexibility offered by this standard at PHY layer and the spectrum sharing options at MAC layer are discussed.

The initial 802.11 standard achieves a rate of 1 Mbps and 2 Mbps in the 2.4 GHz band, while the 802.11b enhances this to support 5.5 Mbps and 11 Mbps. The 802.11a and 802.11g extensions allow up to 54 Mbps. The most recently standardized physical layer is the 802.11n, which supports bitrates up to 600 Mbps thanks to the use of advanced techniques including channel bonding, the use of multiple spatial streams, aggressive coding, short guard intervals and optimized usage of all subcarriers.

The next significant update of the technology will be the 802.11ac standard, adding even more advanced modulation and coding, up to 8 spatial streams and channel bonding up to 160 MHz. With this continuing expansion of the number of streams, channels bonded and modulation and code rates supported, the flexibility offered by the standard is increasing with each generation.

The MAC layer of the 802.11 standard only focuses on the channel coordination functions. In the baseline IEEE 802.11 MAC protocol description, a distributed channel coordination function and a centrally controlled point coordination function are specified. Both however do not provide sufficient QoS guarantees. QoS extensions are specified in the 802.11e standard.

The flexibility offered by the WLAN standard will be exploited to illustrate concepts for smart and cognitive radio design in this book. In Chap. 6 we focus mainly on the physical layer flexibility offered by the 802.11a standard. Considering the modem, we rely on the implementation proposed in [21] with additional flexibility. This will already result in significant energy and performance improvements when the flexibility is exploited smartly. For newer generation devices, the possible gains are higher, at the cost of increased control complexity. Next, in Chap. 7 we focus on the flexibility of the MAC offered for the spectrum sharing. The MAC options considered are applicable to the 802.11a/b/g/n standards, and most likely also the new 802.11ac.

2.2.1.1 The IEEE 802.11a Physical Layer

The 802.11a physical layer has been introduced to achieve bitrates of up to 54 Mbps in the 5 GHz bands. The modulation scheme used for 802.11a is OFDM, where

Table 2.2 802.11a transmission rates

PHY mode	Modulation N_{Mod}	Code Rate B_c	Data Rate (Mbps)
1	BPSK	1/2	6
2	BPSK	3/4	9
3	QPSK	1/2	12
4	QPSK	3/4	18
5	16-QAM	1/2	24
6	16-QAM	3/4	36
7	64-QAM	2/3	48
8	64-QAM	3/4	54

information is transmitted via multiple orthogonal subcarriers. For 802.11a, 48 subcarriers and 4 pilot subcarriers are used, covering a total of 20 MHz in bandwidth including zero carriers. This modulation scheme is very effective for broadband, hence high-speed, transmission since it efficiently handles the frequency selectivity of the channel. Moreover, it can be implemented elegantly using Fast Fourier Transform (FFT) and Inverse FFT (IFFT) digital blocks.

Eight transmission rates have been standardized (Table 2.2), making transmission rate a possible control dimension. A transmission rate is set by configuring both the symbol modulation (N_{Mod}) and the Forward Error Correction (FEC) code rate (B_c). The current value chosen is communicated in the `RATE` field in the Physical Layer Convergence Protocol (PLCP) header of the PHY Protocol Data Unit (PPDU).

Finally, the output power P_{Tx} is limited to 30 dBm or 1 W. The output power can be considered as a PHY control dimension, next to the already introduced digital PHY control dimensions modulation and code rate. As the output power is a result of the Power Amplifier (PA) settings, it is referred to as an analog or front-end PHY control dimension in the remainder of this book. In Chap. 6, we model the impact of those control dimensions on the effective energy and performance, assuming a realistic implementation of the analog and digital modem. However, the output power also has implications on the channel access of a CR, as is discussed in Chap. 7.

2.2.1.2 The IEEE 802.11n Physical Layer

The IEEE 802.11n standard supports data rates up to 600 Mbps. This is mainly achieved using Multiple-Input-Multiple-Output (MIMO) techniques, where up to 4 spatial streams are used in parallel. MIMO techniques that are supported by the standards are space division multiplexing (SDM), transmitter beam-forming and space-time block coding (STBC). The use of MIMO systems mainly increases the throughput and coverage as compared to single antenna systems. Clearly, the possibility to vary the MIMO scheme as function of throughput requirements and channel condition can be considered to be a new flexibility offered by the 802.11n PHY.

Next, the 802.11n offers the possibility to use up to 2 802.11a channels in parallel, which is referred to as channel bonding. As a result, the bandwidth can vary between 20 and 40 MHz, which can be considered to be another new flexibility option. To increase the spectral efficiency even further, also at MAC level, several optimization features have been added, such as frame aggregation or block acknowledgments. Also, to improve the efficiency of the physical layer, a short guard interval option is included, as well as advanced coding such as code rates up to $\frac{5}{6}$ or the option to use low density parity check (LDPC) coding.

2.2.1.3 The IEEE 802.11ac Physical Layer

The IEEE802.11ac Task Group has recently been formed within the IEEE802.11 Working Group. Its goal is to draft a standard for "Very High Throughput" WLANs, with performances significantly higher than those of the IEEE802.11n standard. Because the task group TGac has just started its activity, the standard is not yet available. TGac is currently drafting the detailed functional requirements. The trend of 802.11n towards more flexibility in the MIMO options, channel bonding and more spectrally efficient physical layer options will however persist in the newest standard. More specifically, bandwidths up to 80 or 160 MHz will be supported. Also, advanced multi-stream operation supporting up to 8 streams that can be destined for multiple users is included in the specs. Finally, constellations up to 256 QAM are being discussed.

2.2.1.4 Multiple Access Through Collision Avoidance with Carrier Sensing

We introduce the terminology and communication principles that are standardized in the 802.11 Medium Access Control (MAC) baseline protocol description. According to the 802.11 standard, wireless networks can be arbitrarily complex, and consist of a hierarchy of basic communication networks. A basic wireless communication network is referred to as a *Basic Service Set* (BSS) which is a set of stations controlled by a single *Coordination Function* (CF). The CF manages the access to the radio channel. The *Distributed Coordination Function* (DCF) is mandatory for each BSS and is described first. It standardizes a technique for distributed contention-based channel access. The *Point Coordination Function* (PCF) for centralized contention-free channel access is optional. The PCF operation relies on the DCF. To improve the QoS that can be achieved on top of 802.11 WLAN systems, the *Hybrid Coordination Function* (HCF) has been introduced to improve the baseline DCF and PCF operations.

Further, a BSS can operate in Infrastructure mode (also referred to as regular BSS), with an Access Point (AP) and multiple Stations (STAs). In this scenario, the AP works as a bridge and every transmission is between AP and STA(s). Next, the STAs can operate in Ad-Hoc mode, in which case the BSS is referred to as Independent BSS (IBSS). In this scenario, there are multiple STAs, and no AP. Each STA can be a router to connect the ad-hoc network to the wireline network.

The mandatory 802.11 DCF relies on a Carrier Sense Multiple Access with Collision Avoidance (CSMA/CA) mechanism. When a packet (i.e., MAC frame) is to be transmitted, the channel access procedure should be carried out. This channel access relies on a constant sensing of the channel. The channel needs to be sensed idle for a well-specified duration before the packet can be transmitted.

First, before each packet transmission, an Interframe Space (IFS) needs to be considered. Then, to prevent STAs from accessing the channel simultaneously, a random binary exponential backoff counter is introduced. Two different types of channel or carrier sensing are standardized. The physical carrier-sense is provided by the PHY, and depends on the PHY Clear Channel Assessment (CCA) procedure. Further, a virtual carrier-sensing mechanism is provided by the MAC via a Network Allocation Vector (NAV) counter. This counter can be set since each frame carries the `Duration` field in the header. Hence, any correctly received frame can result in an update of that NAV counter, if the new NAV is larger than the current setting. This NAV overrides the physical layer carrier sensing.

Wireless communication is sensitive to the so-called hidden node problem. It can be seen in Fig. 2.2(a) that STA1 and STA3 can hear STA2, but they cannot hear each other. As a result, their physical layer carrier sensing and their virtual carrier sensing mechanisms cannot detect ongoing transmissions from each other. However, since the interference range is typically at least as large as the transmission (and approximately the carrier sense) range, they can interfere with, hence corrupt, ongoing transmissions of each other. To prevent this, the collision avoidance mechanism has been proposed (Figs. 2.2(b) and 2.2(c)). It consists of two short control frames that are transmitted before each data frame transmission. The first short control frame is referred to as Request To Send (RTS), and is sent by the node that wants to transmit a data frame (Fig. 2.2(b)). If the destination node received the RTS correctly, it replies with a Clear To Send (CTS) message to inform the source it can send its data frame (Fig. 2.2(c)). In the header, the total duration of the planned transaction is listed. STA3, overhearing the CTS, can then set its NAV and delays channel access for the total transaction duration. As a result, the hidden node problem is solved. However, this RTS/CTS exchange results in a considerable additional protocol overhead for each data frame to be transmitted. As a result, it is not mandatory, and is often dropped for short packet transmissions, more specifically for packets that are smaller than the `RTS Threshold`.

It is clear that the hidden node problem can be solved by using a much lower CCA threshold, or alternatively, transmitting the packets at a much higher power. The impact of lowering the sensing threshold or the transmission power on the spatial reuse is a complex problem and not well understood. It is very difficult to derive analytical models to predict the performance of a network for a given setting of those thresholds, since the exact performance depends on a lot of parameters such as traffic density, exact topology etc. As a result, the optimal setting can only be determined at run-time when more information about the exact operating conditions is available. This will be illustrated in Chap. 7.

On top of the DCF, an optional PCF can be used. It enables a poll-and-response MAC for nearly isochronous[1] service. It can be used in the infrastructure BSS only,

[1] Isochronous refers to processes where data must be delivered within certain time constraints.

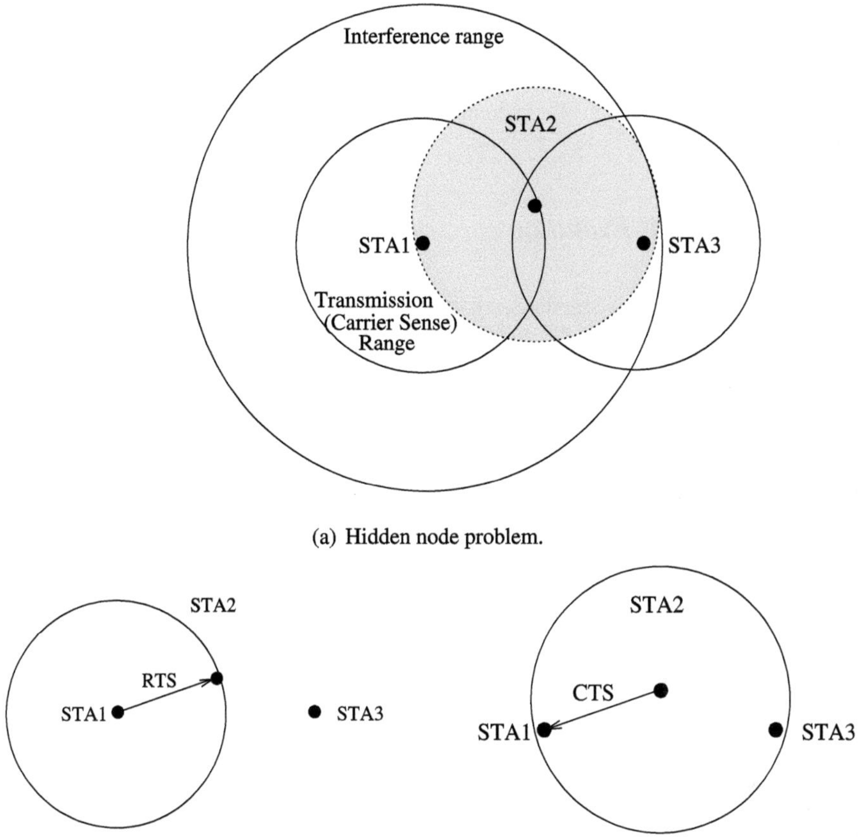

(a) Hidden node problem.

(b) RTS frame exchange.

(c) CTS frame exchange.

Fig. 2.2 The hidden node problem in wireless communications and the RTS/CTS collision avoidance to solve it

since the Point Coordinator (PC) resides in the AP. When the PCF is deployed, channel access is organized in alternating Contention-Free Periods (CFP) and Contention Periods (CP). The PCF suffers from unpredictable, and potentially very large, frame transmission times. Once a station gets access to the channel, it can send a packet up to 2312 Bytes. Assuming that the slowest transmission rate is used, this can take up to 3 ms which is long. The lack of QoS guarantees on top of DCF and PCF has prohibited the use of delay-sensitive applications such as Voice over IP (VoIP) and video conferencing on top of WLAN. To improve on this, the 802.11e has been introduced. One of the mechanisms it includes to improve the QoS is the Transmit Opportunity (TXOP), which is an interval of time when a STA has the right to initiate transmissions. Once a TXOP is acquired, multiple (short) frames can be transmitted. Further, if the TXOP is too small for the current transmission (e.g., since the transmission rate chosen is low), fragmentation needs to be carried out.

This is more fair in networks with varying transmission rates and packet sizes. As long as the TXOP constraint is met, the smart radio can locally optimize it's channel access. In Chap. 6 it will be shown how this can be achieved on top of an 802.11a PHY where the HCF Controlled Channel Access of IEEE 802.11e is used.

2.2.2 3GPP-LTE Evolutions

The 3GPP organization standardizes successive evolutions of cellular standards. This includes first 2G and 2G+ systems, which are were specified in various releases from 3GPP, up to release 7.

Long Term Evolution (LTE) is the latest standard in the 3GPP cellular standards family. It is often referred to as a 4G standard, but it does not yet meat the IMT advanced requirements for 4G since for instance the peak data rate requirement of 1 Gbps is not met. LTE is actually a 3.9G technology, a pre-4G standard and a very important step towards LTE-advanced which is then the true 4G cellular standard. The work on LTE advanced has started mid-2008 focusing of 4G solutions and it is assumed to be standardized in the 3GPP specification Release 10 [22]. The first publicly available LTE service was deployed in December 2009 in Scandinavia. Denmark followed in 2010, while the mobile operators in the US announced a first roll-out only for 2011.

In this section, focus will be mainly on the LTE Rel 8, or the 3.9G of LTE. Then, it will be summarized how LTE advanced builds on this technology.

2.2.2.1 The 3GPP-LTE Air Interface

LTE has ambitious goals with respect to data rates, coverage, latency, operation costs, multi-antenna support, flexibility and seamless integration. Specifically, the goal is to achieve 100 Mbps in the downlink (DL) and 50 Mbps in the uplink (UL).

These peak data rates are based on the assumption that a single user is allocated the whole bandwidth with the highest data rate modulation during a transmission time interval (TTI = 1 ms). With different MIMO configurations ($2 \times 2, 4 \times 4$), these peak data rates increase with the number of parallel streams transmitted. However, when the number of active users in a cell increase, these rates will decrease.

LTE supports different multiple-antenna modes. The baseline configuration of the number of antennas for MIMO is 2×2. Also, 4×2 and 4×4 configurations can be considered. In the DL, the supported modes are: spatial division multiplexing (SDM), beamforming, and single-stream transmit diversity mode(s). In the transmit diversity modes each antenna transmits the same information but possibly with different coding. Receive diversity is mandatory in LTE. A typical example is maximum ratio combining. These modes depend on the capabilities of the UE and the number of receive antennas and should be adapted slowly to reduce control overhead (at the beginning of the communication or at most every 100 ms). Thus switching between MIMO modes is possible depending on channel conditions. In the UL, MIMO schemes should be adapted to reduce complexity in the terminal. The baseline is to have one transmit antenna to save cost and battery power. MU-MIMO can be used. Multiple user terminals may transmit simultaneously on the same resource block through spatial division multiple access (SDMA). The UEs sharing the same resource block have to apply mutually orthogonal pilot patterns [23].

The physical layer of LTE uses Orthogonal Frequency Division Multiple Access (OFDMA) in the DL and Single Carrier–Frequency Division Multiple Access (SC-FDMA) in the UL. OFDMA is a practical multiuser technique that allows data to be sent to multiple users on a subcarrier-by-subcarrier basis. On the other hand, SC-FDMA, being a single subcarrier system, offers the advantage of reducing the peak-to-average-power-ratio (PAPR), hence increasing the transmitter RF power amplifier (PA) efficiency and reducing the power consumption of the terminal or User Equipment (UE). The main difference between SC-FDMA and OFDMA is a DFT processing block. In an SC-FDMA signal, each subcarrier used for transmission contains information of all transmitted modulation symbols since the data has been spread by the DFT over all the subcarriers. Oppositely, in OFDMA, each subcarrier carries information from one modulation symbol [24].

LTE also allows for flexible bandwidths going from 1.4 MHz to 20 MHz [25]. The possibilities are: 1.4 MHz, 3 MHz, 5 MHz, 10 MHz, 15 MHz, 20 MHz. It applies to both FDD and TDD. Half Duplex FDD is also supported. This allows operators to provide a whole suite of new services with different bandwidth allocations, or in addition makes it possible to deploy LTE with high peak data rates even when the spectrum availability does not allow for 20 MHz bands. Smaller bandwidths allow for a more flexible and even opportunistic frequency allocation.

Note that the transmission bandwidth is slightly smaller than the channel bandwidth to allow a frequency separation from other bands. The sub-carrier spacing is constant (15 kHz) regardless of the transmission bandwidth. To allow for operation in differently sized spectrum allocations, the transmission bandwidth is instead varied by varying the number of PRBs which will be explained in Sect. 2.2.2.3.

In terms of mobility, LTE is optimized for low speeds (0–15 km/h). A high performance should be achieved between 15 and 120 km/h. Mobility across the cellular network should be maintained between 120 and 350 km/h (up to 500 km/h depending on the frequency band) [26].

2.2.2.2 Multiple Access in 3GPP-LTE

In this section, it is explained how the resources are used among users for a single LTE cell, and among LTE cells. First, the OFDMA resource allocation is explained, which results in a lot of degrees of freedom and hence opportunities to adapt to the run time channel and load conditions of the different users. Next, frequency planning is introduced. It is illustrated that an important trend exists to go from a design time planning of the network toward a distributed planning of the network during the run time, taking into account information that is only present during

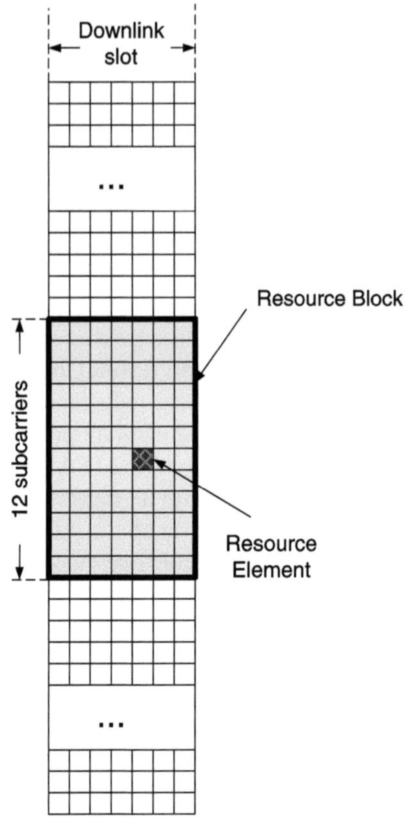

Fig. 2.3 Flexible resource allocation in time and frequency for LTE

the run time. Finally, handover between cells is briefly summarized, which is an additional flexibility present in current wireless networks.

LTE uses a time-frequency-space grid to allocate resources to users in the DL and in the UL (Fig. 2.3). Users are allocated a certain number of resource blocks consisting of 12 consecutive subcarriers during a time slot of 0.5 ms. In the time domain, DL and UL transmissions are organized in radio frames of 10 ms, both for TDD and FDD.

The same coding and modulation is applied to all groups of resource blocks belonging to the same L2 PDU (packets at MAC level) scheduled to one user within one TTI (subframe) and within a single stream. Therefore, the smallest granularity for assigning different modulations or profiles is 12 consecutive subcarriers during a 1 ms period.

In total, each resource block (PRB) consists of 84 resource elements (12 subcarriers during 7 OFDM symbols). However, some of the resource elements within a resource block will not be available as they are already occupied for reference signals or control information.

Even though a downlink resource block is defined in terms of PRBs the downlink resource block assignment is carried out in terms of pairs of PRBs, where each pair

Table 2.3 Transmission bandwidths in LTE

Channel bandwidth [MHz]	1.4	3	5	10	15	20
Number of resource blocks [NRB]	6	15	25	50	75	100

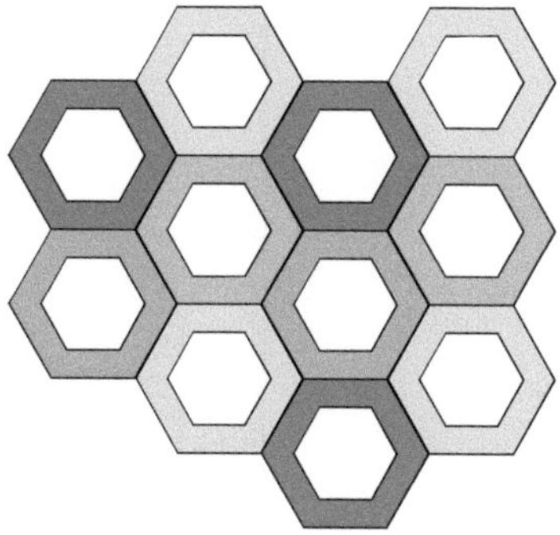

Fig. 2.4 Frequency planning for inter-cell interference (*white* denotes the central region and the *shaded regions* have a less efficient frequency reuse)

consists of two, in the time domain, consecutive resource blocks within a subframe. Additionally, QPSK is used for all control information and data can be transmitted with different modulations like QPSK, 16-QAM, and 64-QAM.

Thanks to the flexible resource allocation in time, frequency and space, and the scalable bandwidths supported by LTE, it is expected that at least 200 users per cell should be supported in the active state for a bandwidth of up to 5 MHz and at least 400 users for higher bandwidths. More users are expected to be supported in a passive (sleeping) state [26].

In a 20 MHz bandwidth, which is the maximum allowed bandwidth, up to 100 Resource Blocks (RB) could be allocated to different users (Table 2.3). A RB is the minimum frequency-time unit allocated to a single user and consists of 12 sub-carriers. Assuming 400 active users per cell should be supported, each user should be allocated 1 RB every 4 consecutive sub-frames (1 ms TTI). The throughput per user in this worst-case scenario will be around 250 kbps. This throughput is for the downlink (DL) direction and will slightly decrease when considering the overhead of the control information.

The cell-edge performance is a more challenging task since inter-cell interference can degrade the system capacity. A way to limit this interference is through inter-cell interference coordination, which can be achieved by restricting the transmission power in different parts of the spectrum in combination with a flexible frequency planning. A possible flexible frequency planning is a soft frequency reuse scheme which is characterized by a frequency reuse factor 1 in the central region of a cell, and a less efficient frequency reuse near the cell edge as in Fig. 2.4.

When the mobile station (or UE) is near the base station (or eNB), the received power of the user signal is strong, and the interference from other cells is weak. So at the inner part of the cell, all the sub-carriers can be used to achieve high data rate communication (Fig. 2.5).

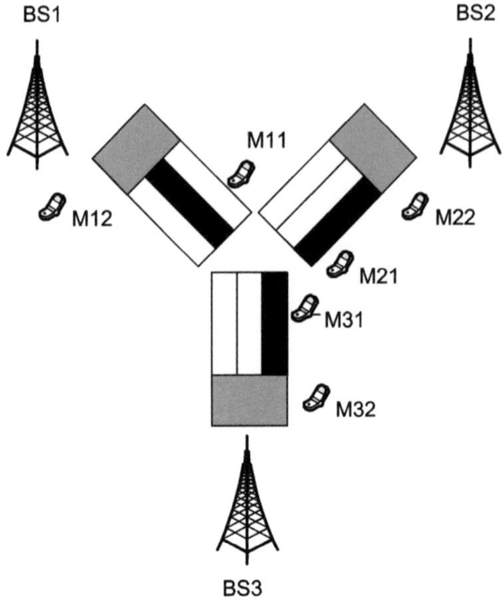

Fig. 2.5 Frequency planning for inter-cell interference (the entire band is used in all cells for the central regions)

In Fig. 2.5, mobile stations 11 and 12 are connected with base station 1, while mobile stations 21 and 22 are connected with base station 2, and mobile stations 31 and 32 are connected with base station 3. It can be noticed that mobile stations 11, 21 and 31 are located at the intersection of 3 cells, while mobile stations 12, 22 and 32 are in the center of their respective cells. For the three mobile stations at the cell edge, different sub-carriers are allocated to avoid the co-channel interference. For the mobile stations near the base station, all the sub-carriers can be allocated and a lower output power can be used, to achieve a high frequency reuse. In this example, the frequency reuse factor is 3 for the cell edge and 1 for the inner part of the cell. For the inner part of the cell, through the limitation of the transmission power, some isolated islands are formed that do not interference each other. To conclude, the benefits of using this flexible frequency planning with soft frequency reuse scheme are:

- High bitrate at the cell center;
- Less interference at the cell edge, making easier the channel estimation, synchronization, and cell selection;
- Overall improved spectral efficiency of the network.

To achieve the benefits of soft frequency reuse, a tight coupling between frequency allocation and output power is needed. This can be achieved by a thorough network planning carried out during the roll out phase of the network. However, with the trend towards smaller and smaller cells, a more distributed approach towards power and frequency planning is needed. In recent years, Self-management (Self-X) technologies that fully automate the tasks of managing (i.e. configuring, monitoring, and optimizing) a cellular network are emerging as an important tool in reducing costs for the service provider and will be a distinguishing feature of

LTE networks. In [27] this Self-X technology is used for the self-configuration of fractional frequency reuse for LTE that uses only local information. Their solution achieves a trade-off between locality of the information, optimality and stability of the solution. This trend to move from design time planning of the network towards self-configuration during the run time, taking into account specific information that can only be obtained during the run time, is the basis of the design methods proposed in this book.

2.2.2.3 LTE-Advanced

LTE advanced is the push towards 4G wireless technologies. It aims to even further data rates than LTE [24] in line with 4G requirements. The most important requirement is the very high data rate, which is 100 Mbps for high mobility and as high as 1 Gbps for low mobility. This low mobility mode is the focus, since LTE advanced mainly aims at achieving high data rates. However, mobility support across the cellular network up to 350 km/h (even up to 500 km/h depending on the frequency band) is also included.

To achieve 1 Gbps, the bandwidth is increased up to 100 MHz. For backwards compatibility with LTE, this is implemented by aggregating up to 5 carriers with a bandwidth of 20 MHz. In addition, the number of spatial streams is increased up to 8×8 for DL and 4×4 for UL. Next to SDM, the MIMO modes include beamforming to improve the performance at the cell edges. For this, multi-cell beamforming is also possible, which means that multiple eNB's cooperate to transmit data to a cell edge user. In addition to cooperative multipoint transmissions, cell edge performance can be improved by means of relaying. Both layer 1 and layer 3 relaying is considered. The target spectrum efficiency is 30 bps/Hz for the downlink and 15 bps/Hz for the uplink. Finally, the number of active users that should be supported in a 5 MHz band is increased to 300.

To make optimum use of all the possible frequency bands, and with a flexible bandwidth up to 100 MHz, it is clear that the frequency planning and configuration is even more flexible than for LTE. In addition to that, with relaying and multi-cell beamforming, the number of possible transmission modes is increased dramatically. As a result, the trend towards self-configuration and self-optimization is supported even more for LTE-advanced. Both flexible spectrum usage and more Self-X are important features of cognitive radio networking.

2.3 Spectrum Access Flexibility

Spectrum is becoming a major resource constraint when designing radio systems. The accelerated deployment of broadband personal communication coupled with the continuously increasing demand for large data rates results in an increasing spectrum scarcity. Wireless technology on the other hand is becoming more and more flexible so that a more opportunistic use of the spectrum or air interface becomes possible.

The need for, and possibility of, a more flexible use of the spectrum is currently recognized and investigated by the major regulatory/standardization bodies (IEEE, ITU, ETSI, FCC, EC) as well as many national spectrum administrations. Worldwide, frequency bands are being opened up for more flexible access. One well-known example of this, is the debate on flexible re-allocation of the so-called 'digital dividend' and 'TV white spaces' which is regularly grabbing headlines in the media. A recent decision by the FCC to open up unused TV spectrum for unlicensed use is fueling this debate even more. In Europe too, there is a call for more ambitious use of the free spectrum space (e.g. through the WAPECS initiative).

In addition to this new trend towards an opportunistic use of spectrum by secondary users, the coexistence problem in the unlicensed bands in a well-known problem that is not trivially solved. In this section, we discuss coexistence problems, both in the case of secondary and primary user coexistence in the TV white spaces or licensed bands, and coexistence in the unlicensed bands. In this book, it will be illustrated how to design more smart or cognitive radios based on a design example for both coexistence scenarios. In Chap. 5 it will be shown how to improve coexistence in the ISM bands. In Chap. 4 it will be shown how an opportunistic radio can avoid interference with a licensed technology.

2.3.1 The ISM Band: Coexistence in Unlicensed Bands

Spectrum sharing in the unlicensed bands is based on a best effort principle, i.e., nodes are equipped with some mechanism to guide spectrum access in a distributed way. However, no guarantees about the QoS can be set, and as a result it is commonly believed that spectrum sharing in the unlicensed bands offers a worse QoS to the end user compared to licensed or fixed access. The CR paradigm investigates how more guarantees can be given when the spectrum is to be shared among possibly heterogeneous technologies. This can for instance be achieved by making more optimal use of the existing flexibility options such as power control, rate adaptation or channel selection. In addition, the spectrum access in unlicensed bands is typically guided by means of a Listen-Before-Talk (LBT) etiquette. This means that node have to *sense* the channel before they can access it. A more optimal configuration of the LBT rules is another important configuration knob to control the spectrum access of a cognitive radio.

To explain the coexistence problem when heterogeneous networks with a similar licensing status have to share the same spectrum, focus in this book is on the coexistence between two major wireless standards that operate in the 2.4 GHz ISM band, namely 802.11g Wireless LAN [28] and 802.15.4 Sensor Networks [29]. Their overlapping frequency channels are shown in Fig. 2.6. The characteristics of both networks are very different, resulting in a problem that is asymmetric in nature. Indeed, the output power of 802.15.4 devices is typically as low as 0 dBm [30], whereas the

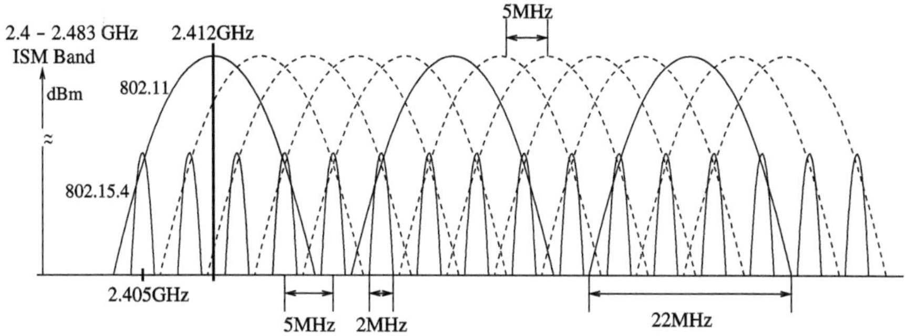

Fig. 2.6 802.11 and 802.15.4 channels in the 2.4 GHz ISM band

output power of 802.11g devices is typically 15 dBm or above. Also, 802.15.4 sensor networks are designed to monitor the environment or buildings, and can be very large, while 802.11 networks are mostly local hotspots organized around an Access Point (AP). Finally, sensor network applications are not demanding in terms of throughput, which means that their duty cycle is relatively low. Because these networks are very different in nature, achieving smooth coexistence between those networks is not trivially achieved since that would require the detection and channel access techniques of both technologies being harmonized. While their operating characteristics are very different, both rely on a LBT channel access mechanism. Nevertheless, the LBT mechanism cannot give any guarantee about the caused interference, and more cognitive adaptation will be needed to improve the coexistence performance.

2.3.1.1 IEEE 802.11h for Spectrum and Transmit Power Management Extensions

Interest in wireless technology has experienced explosive growth over the past decade. Due to the finalization of many standards, the development of wireless applications has eased, which has contributed to increased spectrum use by a variety of heterogeneous devices, standards, and applications. This holds especially true for the Industrial, Scientific and Medical (ISM) bands that are unlicensed and hence host the most heterogeneous mix of networks. To address the coexistence problems in those bands, the IEEE has started the 802.11h Working Group to make recommendations for better future coexistence [31]. It solves problems like interference with satellites and radar using the same 5 GHz frequency band. It was originally designed to address European regulations but is now applicable in many other countries. The standard provides Dynamic Frequency Selection (DFS) and Transmit Power Control (TPC) to the 802.11a MAC that especially coexists with radar systems. Both DFS and TPC will be used further in this book to improve coexistence performance of a CR.

2.3.1.2 IEEE 802.19 Coexistence Technical Advisory Group

The IEEE 802.19 Technical Advisory Group deals with coexistence between unlicensed networks, which is hence applicable to many of the IEEE 802 wireless standards. The first example of wireless coexistence was between IEEE 802.11 and IEEE 802.15.1 (Bluetooth), both operating in the 2.4 GHz ISM band. Coexistence for those networks was first addressed in the IEEE 802.15 Task Group, which produced a Recommended Practice on Coexistence of IEEE 802.11 and Bluetooth. Since March 2006 the IEEE 802.19 TAG addresses coexistence between wireless standards for unlicensed wireless networks under development within IEEE 802. When a new standard (or amendment to a standard) for an unlicensed wireless network is being developed the working group may develop a Coexistence Assurance (CA) document that is reviewed the IEEE 802.19 TAG. At this moment, CA documents exist for IEEE 802.15.3c, IEEE 802.16h, IEEE 802.11y, IEEE 802.11n, IEEE 802.15.4a and IEEE 802.15.4b.

TV bands are generating a lot of fuss and interest within IEEE 802 because working groups like 802.11, 802.15, 802.16 and 802.22 have their eyes on this new (and very good) part of the spectrum. The IEEE 802 Executive Committee approved the formation of a "TV white space coexistence study group" in the 802.19 Coexistence Technical Advisory Group. The Study Group stopped its activity after the March 2009 meeting. IEEE 802.11af is the new study group focusing on a PHY and MAC for the TV bands.

2.3.2 *The TV White Spaces: Spectrum Sharing in Licensed Bands*

The main interest in the TV bands is driven by the worldwide switchover from analogue to digital terrestrial TV which will release a large portion of spectrum known as the "digital dividend", and might also make available the guard bands between existing or new broadcasting channels (the so-called TV White Spaces or TVWS) available for opportunistic use. These frequency bands are very valuable as they provide propagation characteristics that enable signals to penetrate thick walls and travel long distances. The general consensus worldwide is that at least part of this digital dividend and TVWS spectrum should be allocated for mobile broadband services.

The debate on these so-called 'white spaces', is regularly grabbing headlines in the media with a noticeable interest from and role for new players in the wireless field. The debates centers on the free use of White Space Devices (WSD), i.e. devices that could opportunistically make use of free spectrum for various goals. Parties in favor of these devices see major opportunities for the freed up spectrum, including more mobile broadband services, offering better quality of service, to more users. Traditional users of the freed-up spectrum bands and adjacent spectrum bands such as broadcasters are concerned about the consequences of this type of unlicensed use that could interfere with licensed use. They demand a highly regulated approach.

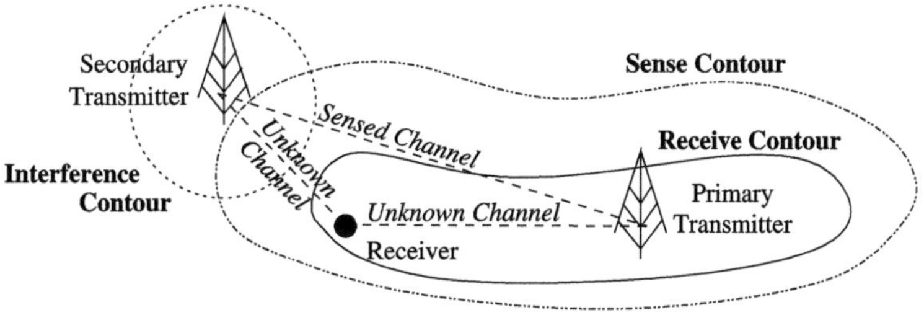

Fig. 2.7 Spatial reuse of the TV white spaces requires large safety margins to ensure that the receive contour of the primary transmitter is protected

In the US, the FCC adopted rules for unlicensed use of television white spaces already in 2008. This announcement was a major endorsement for Cognitive Radio and will softly introduce spectrum sensing technologies in the market in the coming years. The FCC has proposed to use geo-location technology as a primary measure to combat interference: i.e. to check in a database whether a certain frequency band is available or not at a certain location before authorizing the unlicensed use. However, the FCC also stipulates that the geo-location based devices should be complemented with sensing technology:

The Commission also has required that devices include the ability to listen to the airwaves to sense wireless microphones as an additional measure of protection for these devices.

To implement reuse of the TV white spaces, from a platform point of view, the main challenges are related to the sensing requirements. Indeed, as illustrated in Fig. 2.7, it is impossible to sense the exact impact of the secondary transmitter on the primary receiver that should be protected. It is only possible to sense the channel from the primary transmitter to the secondary transmitter, and huge safety margins are added in order to accommodate unknown blocking in the sensed channel. Indeed, if the sensed channel would be blocked, the primary transmitter would seem to be further away than it actually is, and the secondary transmitter could falsely assume that the channel is free. For the TV white spaces, sensing up to −116 dBm is hence targeted, which is well below the noise floor in those bands.

The FCC's conclusion however stated that *devices do not consistently sense TV or wireless microphone signals and the transmitter is capable of causing interference to these signals.*

September 23, 2010, the FCC issued a Second Memorandum Opinion and Order [32] determining the final rules for the use of WSD. The new rules remove the mandatory requirement that WSD should include sensing technology to detect the signals of TV stations and low-power auxiliary service stations (wireless microphones). The FCC states that the geo-location and database access method and other provisions of the rules will provide adequate and reliable protection for incumbent devices, thus making spectrum sensing not necessary since this mandatory requirement would not best serve the public interest. However, the FCC recognized the value of sensing for TVWS in the following statement: *We continue to believe that spectrum sensing will continue to develop and improve. We anticipate that some form of spectrum sensing may very well be included in TVBDs on a voluntary basis for purposes such as determining the quality of each channel relative to real and potential interference sources and enhancing spectrum sharing among TVBDs.*

Although spectrum sensing is no longer a mandatory requirement, the FCC has still defined the technical rules for its use. With regards to the geo-location database, the key specifications are still to be drafted and some key issues clarified, such as how many databases will be in place, who will create and manage them, will they public or closed, how will they be certified, and so on.

In Europe, work on a pan-European specification for cognitive radio systems (CRS) is taking place within the Electronic Communications Committee (ECC) of the European Conference of Postal and Telecommunications Administrations (CEPT). In September 2010, a draft report [33] was released assessing the appropriateness of the sensing and geo-location techniques to provide protection to the existing radio services in the context of a diversified range of envisaged deployment scenarios for WSD. In this report, the spectrum sensing technique (local sensing in which detection is carried out independently by each device) employed in this study is regarded as not reliable enough to guarantee interference protection for DTT receivers and programme-making and special events (PMSE) systems. In the DTT scenario, spectrum sensing does not guarantee a reliable detection of the presence/absence of the broadcasting signals at the distance corresponding to the interference potential of a WSD. Therefore, the use of a geo-location database appears to be more reliable in this scenario. In the case of the PMSE deployment scenarios, temporal fading caused by multipath propagation is likely to be one of the main factors affecting the ability of WSDs to protect PMSE systems from interference. In some cases, this may lead to very low detection threshold, far below the WSD receiver noise floor, making spectrum sensing techniques quite impractical. Thus, the ECC argues that setting sensing thresholds very low in order to protect incumbent services, would result in increasing device cost and complexity as well as a in reduced number of available channels. In its view, this would limit the commercial deployment of WSD and reduce the potential value to end-users. Therefore, the report concludes about the primary need to employ geo-location/database access, since sensing alone would not guarantee interference protection. In case geo-location/database access can provide sufficient protection to the broadcast service, sensing should not be a requirement, since its potential benefit still needs to be further assessed.

2.3.2.1 IEEE 802.22 Wireless Rural Access Networks

The IEEE 802.22 Working Group (WG) was formed in November 2004, after the FCC released its Notice of Proposed Rule Making (NPRM) for the TV bands in May 2004. This WG was specifying an air interface (including PHY and MAC specifications) for Wireless Regional Area Networks (WRAN) to coexist with legacy TV transmission relying on cognitive capability. It was the first standard designed for opportunistic spectrum access, and hence the standard is summarized here. IEEE

802.22 only focuses on fixed devices, while the IEEE 802.22a is to add mobile and portable device functionality to IEEE 802.22. First, the application domain is discussed briefly, to better understand the ultimate goal of the standard and hence the decisions made for spectrum sensing, analysis and decision. The main application target for 802.22 systems is wireless broadband access in rural and remote areas. Because of this specific rural application domain targeting lowly populated areas, the success of the standard was low. However, the sensing features for opportunistic spectrum usage introduced in the standard are interesting and hence briefly summarized in this chapter.

2.3.2.2 System Overview

The use of the lower frequency bands are particularly useful for rural access because of the favorable propagation conditions encountered for those lower frequencies. Although the population density is often very small in rural areas, large coverage areas might render the deployment of 802.22 Base Stations (BSs) a profitable business. These lower frequency bands are licensed for TV broadcasting and Wireless Microphones. However, many TV channels are largely unoccupied in many parts of the United States, and often TV is delivered through cable access or satellite. As a result, opening up those bands for WRAN systems could make a good case, both from business and technical points of view. Next to the main WRAN application domain, 802.22 networks can also be used for smaller markets such as small businesses or home offices. But the main goal is delivering broadband to households in rural areas.

An example of a deployed 802.22 network is given in Fig. 2.8. The 802.22 networks operate in a fixed point-to-multi-point topology where a BS controls a cell consisting of a number of Consumer Premise Equipments (CPEs). The BS is an entity installed by an operator, and controls the cell strictly. Next to more traditional medium access control, that addresses when to transmit, it decides on how CPEs should access the spectrum. Moreover, the BS maintains control of a distributed sensing strategy to keep track of potential primary users (TV or wireless microphone signals). Clearly, it possible to have multiple 802.22 cells that interfere. This is aggravated because of the very large transmission area of those systems. Coexistence issues of 802.22 cells are hence also addressed in the 802.22 standard.

2.3.2.3 Spectrum Sensing

One of the important components of the IEEE 802.22 document to achieve the required cognitive capability is related to spectrum measurements. The spectrum measurement in 802.22 is primarily based on transmitter detection. In order to check the presence of primary signals, 802.22 devices need to be able to detect signals at very low Signal-to-Noise Ratio (SNR) levels. Since the detection is done at low SNR, it is assumed that the detection of TV signals is done in a non-coherent manner, which means that no synchronization is needed [34].

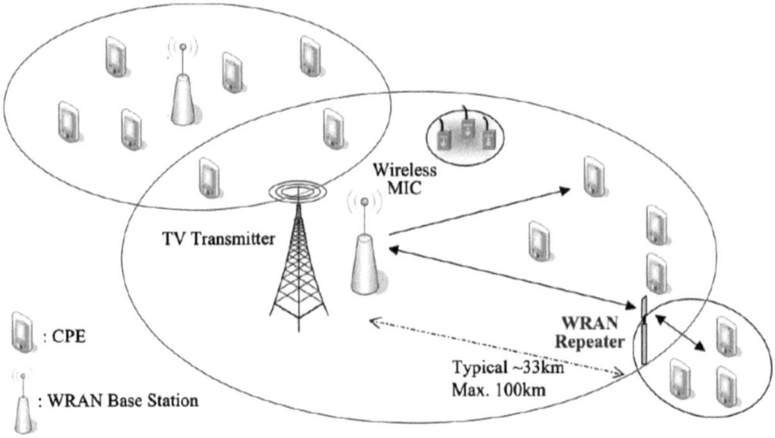

Fig. 2.8 802.22 deployment scenario

The required accuracy of the spectrum sensing, the frequency band and time period, is determined in a centralized way by the BS. Using the local measurements, the BS can establish a spectrum occupancy map. The BS does not require the same sensing accuracy of each CPE, and algorithms to optimize or distribute the sensing load across CPEs can be used. To optimize the sensing, 802.22 devices are supposed to be equipped with a dedicated omni-directional antenna for sensing. This is in addition to a directional antenna which is used for data transmission in the target direction, minimizing the interference area. To be able to optimize the sensing accuracy of the omnidirectional antenna, it would most likely have to be mounted outdoors [34]. 802.22 devices can be instructed to perform in-band or out-of-band sensing, where a band denotes the TV band currently used by the cell. For in-band sensing, the 802.22 communication needs to be temporarily halted, in order not to interfere with the sensing. There clearly is a trade-off between speed at which a primary TV signal can be detected, and the efficiency or throughput achieved by the 802.22 cell. To avoid too frequent long connectivity halts, a two-phase sensing mechanism is proposed (Fig. 2.9). Fast sensing, i.e. based on a simple and fast sensing technique, is performed more frequently. After one (or more) fast sensing periods, the BS can decide whether to perform a fine sensing. This fine sensing takes more time but should in fact only be carried out if the fast sensing results are not sufficient to draw conclusions. Given the fact that TV signals do not come on the air frequently, this two-phase sensing method proves highly effective [34].

If multiple 802.22 cells operate in the same area, it is required that their sensing strategy is synchronized (i.e., they should halt communication when other cells sense). Since coexistence among different 802.22 cells is an important issue, such synchronization is embedded in the 802.22 standard. Contrary to the TV signals detection, sensing of wireless microphone transmissions is much harder as these transmit at a much lower power and occupy much lower bandwidths. Therefore, in addition to transmitter detection, a second sensing option is enabled in the 802.22

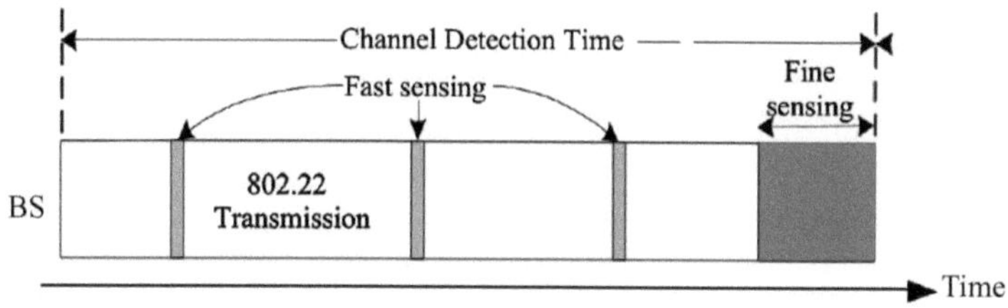

Fig. 2.9 802.22 two-phase in-band sensing

standards. This second option relies on the transmission of beacons by the microphones themselves or a special device carried by microphone operators. This primary network information monitoring is embedded in the 802.22 MAC.

The exact sensing algorithms to be used during the fine and fast sensing periods are not standardized, but an extensive list is provided as annex to the standard.

2.4 Operation Across Technologies: Cognitive Radio

In the previous section, coexistence between technologies was discussed. Optimal sharing is achieved when technologies are not only aware or each others presence, but also act as a function of the coexistence scenario and ideally cooperate. Cooperation between different technologies was first possible in the form of medium independent handover. Although the handover capabilities were merely a responsibility of the terminal, the fact that a standard and information sharing across technologies is present, can be considered as a first example of operation across technologies. Current standards focusing on cognitive radio or dynamic spectrum access networks are also discussed. These standards aim a step further towards the true seamless reconfiguration of multiple terminals and networks. For these future standards, both ETSI and IEEE initiatives are discussed.

2.4.1 Mobile Independent Handover: IEEE 802.21

This standard defines media access independent mechanisms that enable (especially through link layer information) an efficient handover between 802 mobile terminals (capable of supporting different link-layer technologies) and may facilitate the handover between 802 and cellular terminals. The coverage goes from the 802.15 pico-cells and the 802.11 micro-cells up to the 802.15 and 3GPP macro-cells with overlapping coverage. This standard is intended to improve users' experience of mobile devices for 802 heterogeneous networks in a wired or wireless environment and to allow a seamless handover, when the network conditions support it.

The media types that this standard supports include those specified by the IEEE 802 wired and wireless family of standards, as well as those specified by the Third

Generation Project (3GPP), and Third Generation Partnership Project 2 (3GPP2). This standard also addresses the support for handover of mobile and stationary users. As expected, handover for mobile users will happen when changes in the wireless link conditions occur. Not so obvious is the case of stationary users. In general, when one network becomes more attractive to another due to network environment changes, a handover for a stationary user might occur, for example, when a user starts an application that requires higher data rate channel. In any case, the handover should be as transparent as possible to the user.

Furthermore, the IEEE 802.21 standard supports cooperative use of information at the mobile node and within the network infrastructure. This means that both the mobile node and the network can make decisions about connectivity based on measurement reports supplied by the link layer. These reports may include metrics like signal quality, synchronization time differences, and transmission error rates. It is assumed that the mobile terminals and the points of attachment such as the base stations and access points may be multimodal, i.e., capable of supporting multiple radio standards and simultaneous connections on more than one radio interface.

2.4.2 Dynamic Spectrum Access Networks: IEEE DYSPAN

The IEEE 1900P standards committee was established in 2005, in order to promote new standards and technologies for better spectrum management. It has been reorganized as Standards Coordination Committee 41 (SCC41). In December 2010 all activities of SCC41 were transferred to the Communication Society Standards (CSSB). And CSSB established a new committee to manage the SCC41 activities under the new name. IEEE DYSPAN Standards Committee. Two working groups are presented hereunder: 1900.4 and 1900.6, since they standardize the main features of a CR. 1900.4 focuses on the cognitive control and how this should be balanced between the network and the terminal. 1900.6 focuses on the information gathering for a CR.

2.4.2.1 IEEE 1900.4

The task of IEEE 1900.4 working group is to develop a standard defining "Architectural Building Blocks Enabling Network-Device Distributed Decision Making for Optimized Radio Resource Usage in Heterogeneous Radio Access Networks". The objective is to improve the overall composite capacity and the quality-of-service (QoS) of wireless systems in heterogeneous radio access networks (RANs). To achieve this objective the standard defines network resource managers (NRM), terminal resource managers (TRM) and the information to be exchanged between NRM and TRM. This information exchange between network side entities and terminal side entities (network-terminal distributed decision making) is exploited to optimize the radio resource usage including spectrum access control in heterogeneous RANs.

The standard is associated with the heterogeneous wireless environment which includes multiple operators, multiple RANs, multiple radio interfaces and multiple terminals.

NRM and TRM interact with each other and exchange the context information measured at the network and the terminal side. This context information is exploited for the network-terminal distributed optimization of the spectrum usage. The standard considers the reconfigurable base stations and reconfigurable mobile terminals. Any network side radio node is called the base-station which can be a Wi-Fi access pint (AP), a UMTS node-B or a WiMAX base station (BS) etc. The reconfigurable terminals that are considered in the standard may or may not have multi-homing capability, which refers to the capability of a reconfigurable mobile terminal to have more than one simultaneously active connections with RANs.

The standard defines three use cases:

- *Dynamic Spectrum Assignment:* In this use case the frequency bands are dynamically assigned to RANs. It means that the frequency bands are not fixed and can be dynamically changed to be used by RANs and terminals by an operator spectrum manager (OSM).

  Spectrum sharing and spectrum renting among RANs are examples of this use case. In spectrum renting the frequency bands of one RAN is assigned to the other RAN. In spectrum sharing the frequency band of a particular RAN is shared among several RANs. For example a frequency band previously used for WiFi (IEEE 802.11) wireless Internet access can be shared/rented to mobile broadband wireless access WiMAX (IEEE 802.16).
- *Dynamic Spectrum Sharing:* In this use case fixed frequency bands can be dynamically assigned to RANs and terminals. The frequency bands assigned to RANs are fixed but can be shared among several RANs.

  As an example of such use case might be the opportunistic spectrum access of unlicensed secondary systems (IEEE 802.22 or unlicensed WLAN IEEE 802.11 in a possible future CR environments) with the primary licensed but un-utilized part of VHF/UHF spectrum in time or space.
- *Distributed Radio Resource Usage Optimization:* In this use case the decision making is done in a distributed manner by the network and the terminal.

  An example of this use case may be the scenario in which the context of the CWN changes e.g. network load is increased or decreased, change in the channel fading statistics and the availability of a high throughput/user preferable RAN etc. In such a case the TRM analyze the context information and dynamically make its own decisions to optimize the radio resource usage, QoS and its own objectives.

According to the standard there shall be an entity on network side, called Network Reconfiguration Manager (NRM), responsible for managing the CWN and Terminals for network-terminal-distributed optimization of spectrum usage. There shall be an entity on terminal side, called Terminal Reconfiguration Manager (TRM), responsible for managing the terminal for network-terminal-distributed optimization of spectrum usage. The TRM shall manage the terminal within the framework defined by the NRM and in a manner consistent with user's preferences and available context information. This evolution towards policy derivation by the NRM that will guide the TRM do execute it's cognitive control mechanisms in a way that are in line with the overall network objectives is an important enabler for CR. In addition to the decision framework, the entities for information extraction, collection and storage have been standardized. Also for this functionality, it is assumed that TRM and NRM cooperate.

From April 2009, 1900.4 Working Group works on two projects:

- 1900.4a: Standard for Architectural Building Blocks Enabling Network-Device Distributed Decision Making for Optimized Radio Resource Usage in Heterogeneous Wireless Access Networks – Amendment: Architecture and Interfaces for Dynamic Spectrum Access Networks in White Space Frequency Bands. This working group hence focuses on the extension of the 1900.4 standard for the TV White Spaces specifically.
- 1900.4.1: Standard for Interfaces and Protocols Enabling Distributed Decision Making for Optimized Radio Resource Usage in Heterogeneous Wireless Networks. This standard targets the next step with respect to the cognitive control standardization.

2.4.2.2 1900.6

The intended standard defines the information exchange between spectrum sensors and their clients in radio communication systems. The logical interface and supporting data structures used for information exchange are defined abstractly without constraining the sensing technology, client design, or data link between sensor and client. The standard defines a Cognitive Engine (CE), a Data Archive (DA) and a Sensor, and defines all logical interfaces between those entities. This definition is hence in line with various approaches for obtaining information about spectrum use, such as local sensing, distributed sensing or geo-location database. The standardized interfaces are very general to facilitate any sensing hardware, algorithm of architecture. A follow-up project for the 1900.6 working group is under discussion, to see how the standard can be instantiated for more practical sensing approaches or geo-location database.

2.4.3 Reconfigurable Radio Systems: ETSI RSS

Similar to the IEEE DYSPAN initiative, the ETSI standardization is starting to standardize Reconfigurable Radio Systems. RRS is a generic concept to exploit the capabilities of reconfigurable radio and networks for self-adaptation to a dynamically-changing environment with the aim of improving supply chain, equipment and spectrum utilization. The Technical Committee has four working groups on System Aspects, Radio Equipment Architecture, Functional Architecture and Cognitive Pilot

Channel (CPC) and Public Safety. For the hardware flexibility, it is mainly the Radio Equipment Architecture working group that is interesting. About the cognitive control, the Functional Architecture and CPC are the most important initiatives. It is worth to mention that the CPC is an alternative to the sensing or geo-location database approaches. It allows a CR to obtain information about available services and the spectrum use by means of listening to a dedicated channel. The CPC was primarily a European alternative for the CR control and information sharing problem. In [35] it was shown that third-party approaches to facilitate a CR operation clearly have their drawbacks since the current stakeholders in the wireless landscape of course do not like to rely on such a third party for their services. Similar arguments can be considered for the geo-location database that is most likely or possibly also going to require a third party (such as e.g. Google).

Chapter 3
Cognitive Radio Design and Operation: Mastering the Complexity in a Systematic Way

3.1 The Need for a Strategy

Wireless communication networks are complex and dynamic systems. So many variables and uncertainties related to the environment and user requirements exist, which makes it extremely difficult to derive effective static control strategies. As a result, wireless devices need to become increasingly smart by allowing dynamic strategies that are able to track the changes in the network and adapt accordingly. While the trend towards more flexibility both in terms of hardware and policy flexibility was already introduced elaborately in Chap. 1, this chapter focuses on the smart control this implies.

Several challenges need to be addressed when designing efficient control algorithms for these flexible radios. Indeed, next-generation radios and networks need to be *smart* to exploit the opportunity of their flexibility. However, while *dumb* strategies can be readily and efficiently implemented, more complex control strategies pose a burden on the run-time complexity for these radios. Hence, wireless devices need to be smart at two layers: they need to be smart to exploit the dynamics of their environment and they need to be designed in a smart way so that the run-time complexity remains acceptable.

In this chapter, we demonstrate a general control strategy that is able to adapt flexibly to the environment without heavily impacting the run-time complexity. In the subsequent chapters, this strategy is instantiated for case studies to emphasize different aspects of the general framework.

The method proposed in this book can be roughly divided into a design time and a run time process. The degree of intelligence and adaptation enabled in the run time process typically determines how *smart* the radio is. As defined in Chap. 1, a radio is only cognitive when it learns based on feedback from the environment. It will be shown that the run time involves four main tasks: monitoring or observing, determining the current scenario, acting on the scenario following a procedure and finally learning or calibrating the procedure. In the remainder of this book, a case study will then be presented that involves each of the four tasks, but however, emphasis will be on one of the tasks:

S. Pollin et al., *Software Defined Radios*, Signals and Communication Technology, DOI 10.1007/978-94-007-1278-2_3, © Springer Science+Business Media B.V. 2011

- **Observe RT situation:** The challenge in sufficiently observing the RT scenario is the most prevalent in cases where Primary Users (PU) and Secondary Users (SU) have to coexist in an OSA scenario. In Chap. 4 the flow will hence be instantiated in a PU/SU scenario with special focus on the RT monitoring.
- **Map RT Situation to scenario:** Very often several sources of information have to be combined to sufficiently understand the current scenario and act effectively. This will be shown in a case study of IEEE 802.15.4 coexisting with IEEE 802.11, where the goal of the IEEE 802.15.4 network is to find a single channel that is least interfered by IEEE 802.11. Routing layer information, together with PHY layer sensing information, will be combined to enable the IEEE 802.15.4 network to find its single best channel.
- **Execute RT Procedure:** The execution of a well-defined RT procedure does not necessary require a radio to be cognitive. However, it will be shown that the proposed flow can still be used to achieve a run-time efficient design of smart radios that adapt to their environment. The context will be an IEEE 802.11 network in which multiple users coexist and have to share the medium, while meeting performance constraints and minimizing energy cost.
- **Calibrate and learn RT Procedure:** The most important task of a cognitive radio is learning and calibrating how it behaves in the environment. How to achieve this efficiently is illustrated in the context of IEEE 802.11 networks that aim at minimizing the co-channel interference in a distributed way by adapting the output power, sensing threshold and transmission parameters.

This chapter illustrates the generic design flow that will then be instantiated in the next chapters. First, the design landscape is completely described in Sect. 3.2. While Chap. 1 focused mainly on hardware and policy flexibility, it is shown in Sect. 3.3 that the involved challenges for the control solution are even broader. Finally, in Sect. 3.4 the proposed design framework is then introduced.

3.2 The Design Landscape Is No Longer Flat

The purpose of this section is to explain the design landscape for the control of wireless devices and networks. The easiest systems to optimize are those that have limited dynamism.[1] They can be approximated at DT by static models typically designed for the worst case. Hence, relying on these static DT models, the system can be configured completely at DT and does not need to be changed at RT.

However, as we will further detail in this chapter, wireless networks are extremely dynamic and, hence, DT-only frameworks for wireless networks, which always need to be designed based on the worst case operating conditions, are severely suboptimal. Therefore, wireless networks ask for a more elaborate framework that

[1] A system is said to have limited dynamism if $\frac{\text{mean(FoM)}}{\text{max(FoM)}} > (1-\epsilon)$, when using a static configuration that optimizes mean(FoM). FoM denotes the Figure of Merit at RT and ϵ is some small quantity.

tracks the dynamics at RT and adapts to the locally observed changes in the environment. Still, most industrial designers today implement their solutions in a DT-only way due to the reduced manual effort and the limited tool support to go beyond such a DT-only approach. In this book, however, we assume that this isn't the case for the development of wireless devices.

As explained in Chap. 1, next-generation terminals will operate in increasingly complex scenarios, where interoperability and coexistence will become key concerns. The CR paradigm proposes a framework to address these challenges. Due to the increased dynamics and the difficulties faced at DT to model these elaborate coexistence scenarios, Dr. Joseph Mitola III has stated that the ultimate CR would rely on a RT-only framework [15]. Such a RT-only framework would not use a single DT model and, hence, inaccuracies in these models wouldn't propagate to the control procedure used at RT. Furthermore, a RT-only framework would be extremely flexible as each new scenario is learned through experience and interaction with the environment. Unfortunately, the RT algorithms to explore each situation and future evolutions in a holistic way require enormous learning overhead today. Hence, in this chapter, we propose a hybrid DT/RT framework to exploit the potential of the ever increasing flexibility without increasing overhead to unacceptable levels.

We propose to model dynamic systems as a sequence of scenarios. Then DT procedures are built, relying on these DT models. At RT, the RT situation is efficiently monitored and mapped to a specific DT scenario, after which the DT procedure linked with the observed scenario is executed. However, in line with the CR vision, the framework then calibrates these procedures to learn new RT procedures based on interaction with the environment.

3.3 Design Challenges and Opportunities

In this section, we present an overview of the main design challenges and opportunities in the context of wireless networks. In the next sections, we then detail how these challenges are addressed by our proposed framework.

3.3.1 Design Time Complexity

The goal of a smart radio is to adapt its operation to the dynamic environment conditions and user requirements. This is a very challenging task that requires a global approach. Indeed, all layers of a single terminal should be included (going from the flexibility at hardware level all the way to applications running on top of it). In addition, because of the sharing, the interactions of all other users should be considered. Considering the global problem during the design of the smart radio is hence very complex, and a good modeling, verification and testing of the design is a very challenging task. This is a driving force behind the search for cognitive radios, since an ultimate cognitive radio would learn how to adapt during the run-time to the environment and application, making the number of design decisions that should be taken by the engineers significantly smaller.

However, it should be clear that an ultimate software radio does not exist, and the radio operations will always be constrained by the dimensioning of the hardware. While it is possible to design for instance analog SDR solutions that are very flexible (see [36]), the tuning range is of course bounded and discrete. Similarly, for the digital part, it would be too costly in terms of area and power if the SDR would be designed a virtually unbounded number of possible operations and complexity budget. The design of SDR or CR is hence maybe even more challenging compared to the design of traditional ASIC solutions. Indeed, instead of a use case scenario, the designer should be familiar with a broad range of use case scenarios and modes and achieve an optimal design, on average, for the good mix of scenarios and modes. How to determine that good mix, how to trade-off the performance of each individual scenario or mode in that mix, and the run-time choice between different modes and scenarios are all new design questions that did not need to be addressed in traditional designs. Clearly, there is a need for a more systematic design flow, which is the goal of this book/chapter.

Next to the complexity caused by the different modes and scenarios, a large part of the design time complexity is caused by the fact that the problem is inherently multi-dimensional. For instance, a baseband design for SDR is characterized in terms of power, area, performance and flexibility. A better design in one dimension is typically performing worst on the other dimensions. Selecting the optimal trade-off in this multi-dimensional design space is very difficult since it requires balancing standardization, regulation, business aspects and good engineering. None of these aspects can however be omitted when a successful design of a smart radio is targeted.

3.3.2 *The Mountains We Have to Climb*

Wireless networks are complex and dynamic systems. First of all the environment in which they operate is becoming increasingly dynamic, since wireless networks are becoming more popular and, thus, dense. Furthermore, a broad range of applications has to be supported, resulting in heterogeneous performance constraints. In this section, we detail some of the dynamics faced in wireless networks.

3.3.2.1 Channel Dynamics

Wireless channel dynamics are studied extensively in literature [37]. The considered coherence time, i.e., the time duration over which the channel is expected to remain stable with a given probability, can range from the duration of a symbol or bit at physical layer to the scheduling period at medium access control layer [38]. Depending on the exact degree of the dynamics (i.e., coherence time), different adaptation strategies are optimal. For instance, when time variations can be tracked timely, it is optimal to adapt accordingly. However, when variations are too fast, diversity techniques are very powerful. It is however widely accepted that the coherence time for indoor and stationary users can easily be larger than 100 ms (see Chap. 6). This implies that channel variations are fast, but it is still possible to track these variations in the lower layers.

Besides variations in time expressed by the coherence time, channels are typically also varying over frequency. This variation is captured in the coherence bandwidth, and frequency-adaptive techniques exist [37] in case the communication channel bandwidth is larger than the coherence bandwidth.

Finally, the channel is varying in space. This last variation means that different users in the same network often see a very different channel since they are located in a slightly different spot. Furthermore, the use of multiple antennas on a single device exactly exploits the fact that both antennas see a different channel since they are separated in space [39]. Variations over space will be exploited to improve the monitoring in Chap. 4 and to make the RT procedure more energy efficient in Chap. 6. Solutions should hence be robust to the channel variations in time, frequency and space.

3.3.2.2 Application Dynamics

Networked applications communicate in packet bursts that are transmitted over the communication network. Such bursty behavior is typically more difficult to track and to predict compared to more continuous variations. Next to packet burstiness within application streams, the average bitrate requirements can differ a lot from application to application.

Different applications are also sensitive to different performance aspects. Real-time streaming applications prefer strict delay guarantees for a frame, while file transfer is best served by maximizing the average throughput, hence total file download delay. Sensor networking applications are typically not throughput intensive. For those applications, delay and robustness against failure are the most important performance aspects.

3.3.2.3 Network Dynamics

Mobile networks do not necessarily rely on an extended infrastructure to operate and can be established on the fly, i.e., in an ad-hoc way. Alternatively, in infrastructure mode, the network changes as terminals join and leave the network. As a result, devices can take part in networks of virtually any size or topology, and the number of users in the network will vary over time. Solutions should therefore be able to optimize networks of different size or topology. Network dynamism is limited by users' mobility. A terminal remains connected to a single network for seconds (a phone call, while driving) to several hours (WLAN in the home).

When dynamics are slow, the system can be observed through events. For network dynamics, such an event would for instance be a terminal joining the network. Between events, a system can be approximated as being static.

In Chap. 7 it will be illustrated how to adapt the network parameters to improve coexistence with other networks of the same type. In Chap. 5 the coexistence problem of IEEE 802.15.4 networks with dynamic IEEE 802.11 networks will be considered.

3.3.3 The Sharing Challenge

Throughout this book, the sharing problem is a key driver for the control challenge and, hence, the need for a systematic flow. It is not sufficient to solely consider the impact of a local decision, e.g., decreasing the transmission rate, to evaluate local performance. Thus, the results of the proposed framework in its global environment are considered: a single network consisting of multiple users, different networks sharing a same channel, different heterogeneous sharing the ISM band and finally the case where primary and secondary users coexist.

In this section we briefly discuss three challenges related to sharing in a wireless context: the spectrum policy, multi-user interaction and fairness.

3.3.3.1 The Spectrum Policy

Like water, air or oil, the wireless spectrum is a shared resource. This however does not imply that it should be free! Indeed, the base of all economics is scarcity and we have already shown above that wireless spectrum appears to be very scarce today and is, hence, expensive.

As mentioned in Chap. 1, the apparent spectrum scarcity and the success of the ISM bands has coerced regulatory bodies into defining a roadmap towards more local and organic policies for sharing the wireless spectrum. Today, policies cover large areas and remain static over years. The end goal of the regulatory roadmap is to increase the policy dynamism to days, hours or even minutes. This implies that future terminals will operate in a diverse environment, where they can act amongst equal, superior and inferior terminals from a policy point of view.

3.3.3.2 Multi-user Interaction

Wireless devices and networks operating in the same frequency bands interact. Similar to channel dynamics, multi-user interaction or interference has an impact on the channel quality. However, unlike channel dynamics, interference dynamics are created by the actions of radiating objects in the vicinity (microwave ovens, wireless terminals inside or outside the network, ...). Increased interference results in an increased packet error rate at the physical layer or an increased packet collision rate at the MAC layer. For the user, the competition for the spectrum results in an increased channel access delay or a lowered achieved throughput. The impact of the detailed actions of each other terminal in the network has a huge impact on the achieved performance and is hence the main reason why current radios should adapt, intelligently.

3.3.3.3 Fairness

Wireless spectrum is a shared resource. This, however, does not imply that everybody has the same access to it. For instance, while water is a shared resource, in a remote village in Africa it is generally more difficult to access water than in a European town. Unfairness can however be compensated, for instance by installing better irrigation systems in Africa.

Similarly, some wireless devices may experience good channel conditions, while others are experiencing worse situations. This unfairness can be compensated by allowing users with bad channels to access the spectrum longer. Unfairness can also be caused by a disparity of different application requirements. We have, for instance, shown that energy-sensitive wireless devices can throttle the performance of throughput-sensitive devices when both can access the spectrum equally [19]. Is it fair that energy-sensitive users conserve energy, at the expense of the throughput-sensitive users?

This question is difficult to answer as today fairness criteria[2] remain very ambiguous and local. A strong debate is ongoing to find a unifying fairness definition. Only when this debate closes and such a definition is found, can fairness be truly regarded as an objective.

3.3.4 Run-Time Complexity

We have shown that the global design problem is very challenging. However, the goal remains to design an efficient run-time procedure for a radio that can be executed on a handheld, mobile smart or cognitive radio. Wireless communication devices are typically small and battery-operated devices. Hence, algorithms designed for these devices should be run with limited computation resources. Combining the inherent complexity with limited capabilities is the last challenge that should be addressed. The focus of this book is mainly on the run-time operation, that consists of four phases as already introduced in the introduction of this chapter: observing the RT situation, determining the scenario, executing the RT procedure and finally

---

[2] For instance, in wireless networks a plethora of fairness indices are currently being used. We refer interested readers to [40, 41].

calibrating or learning the RT procedure. The full framework, including design time and run time phases, is introduced next.

3.4 Proposed Control Framework

The ultimate CR is built on a RT-only framework and, hence, does not rely on any DT model. As a result, modeling inaccuracies will not propagate to the control procedure at RT. Furthermore, a RT-only framework will be extremely flexible as each scenario is learned through experience and interaction with the environment. Unfortunately, RT algorithms to explore each situation and future evolutions in a holistic way require enormous overhead. On the other hand, DT-only frameworks based on worst-case operating conditions, are known to be severely suboptimal for dynamic wireless networks.

In this book, we present a framework that can exploit the flexibility, while maintaining an acceptable complexity level. This framework relies on design control procedures at DT, that are efficiently executed at RT. This allows for an effective and efficient RT operation. However, as the control procedures are built relying on DT models, inaccuracies in these DT models may propagate to the control procedure. Furthermore, creating DT models is becoming increasingly complex as wireless terminals are facing very complex environments. In line with the CR paradigm, we also embed a learning module to compensate these inaccuracies and to find suitable operating points even in unforeseen situations.

This section is structured as follows. First, we detail the general design concepts in Sect. 3.4.1. In Sect. 3.4.2, we describe the design flow. Finally, the RT operation is detailed in Sect. 3.4.3.

3.4.1 General Design Concepts

In this section, we describe the general design concepts of our framework. In Fig. 3.1, a condensed version of our proposed framework is shown. First, we describe the impact of the increased flexibility provided by the SDR and OSA paradigms on our control framework. Then, we discuss environment awareness and why it is critical in the development of future CR systems. Finally, we discuss how we can learn from extended interaction with the environment.

3.4.1.1 Control Dimensions

The quest for the ultimate CR is driven by new paradigms that increase the HW flexibility (SDR) en the policy flexibility (OSA). As a result, the set of RT adjustable control dimensions (also referred to as *knobs*) of a wireless terminal is steadily increasing. It is important to rely on this knob set to reconfigure the system at RT. The

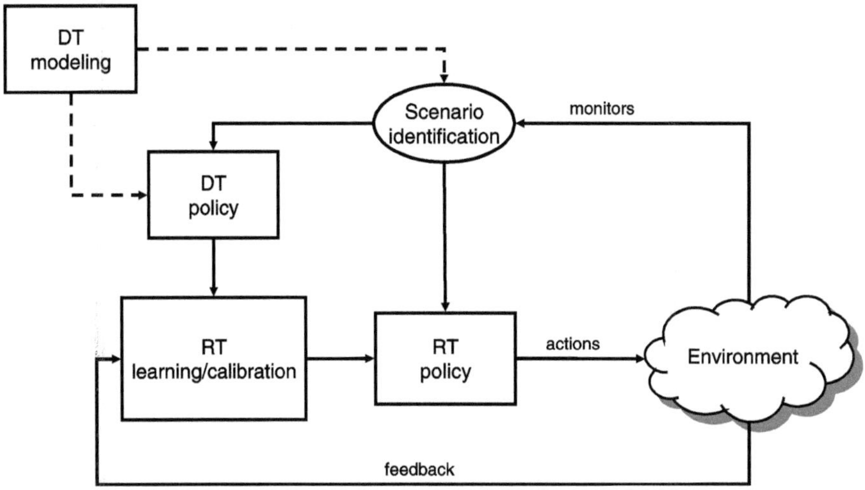

Fig. 3.1 A condensed version of the proposed framework. DT procedures that are established from DT models observe the RT situations and map this to a scenario. The DT procedure linked with the observed scenario is calibrated at RT by the RT learning engine

goal of the proposed framework is to dynamically select the configuration of those knobs to address the user's (or operator's) needs. The more knobs available to effectively control the system, the more gains can be achieved when using the presented framework.

In this book, it is assumed that knobs are available (see Chap. 1). As we are concerned with developing the CR, a control layer, the development of protocols or implementation techniques that make those knobs available are outside the scope of this work.

3.4.1.2 Environment Awareness

Both Haykin and Mitola identified environment awareness as a key element for future CR systems. Indeed, a wireless device today does not operate independently but acts within a larger environment. Awareness of this environment does not imply you can understand it, which is the ultimate goal of a CR. However, the opposite does hold: understanding an environment can't be achieved without closely monitoring it!

When a wireless device monitors the environment, it can map observations to an environment model, created at DT. This presumes that the environment can be at least partly analyzed and modeled at DT. Such a model should however be very parameterized and even abstracted to a high level. This is usually feasible as we show in subsequent case study chapters.

In this way, we avoid having to do everything at RT (which provides too much RT execution overhead) or everything at DT (which is too complex in practice or involves worst case design margins). Based on the hybrid information, the CR can then synthesize the combined model of the environment at RT.

3.4.1.3 Efficient and Effective Calibration at Run-Time

Especially in best-effort systems or when system constraints are soft, wireless terminals are allowed to do even more than passively monitor and track the environment. Based on feedback from the environment, they can explore and learn the behavior and response of the environment, while calibrating the DT control procedures.

The essential information, needed to learn and calibrate, is the feedback from the environment. While monitoring is passive, feedback concerns action (from the terminal) and reaction (from the environment). A CR can only truly start understanding its environment, when it is able to interact and observe the environment response. Unsurprisingly, Haykin stated: "... therefore the CR is, by necessity, an example of a feedback communication system" [17].

Exploration and learning allow to compensate for difficulties faced at DT modeling. As we explain below, at DT we cluster different RT situations into system scenarios. For each system scenario, a DT procedure is then designed. By calibrating the DT procedure for the observed scenario, we customize the DT procedure towards the observed RT situation and, hence, further gains can be achieved.

3.4.2 Design-Time Flow

In this section, we present the different steps to be taken at DT. As illustrated in Fig. 3.2, the first step is modeling. After this, control dimensions of the system and environment dynamics are identified. These control settings are then mapped onto the objective space. Next, we cluster the space into a small set of system scenarios and define rules how these system scenarios can be identified at RT. The final result of the DT flow is a DT procedure for each system scenario. Below, we explain these steps in further detail.

3.4.2.1 Design-Time Modeling①

Any DT procedure can only be seeded by what is known at DT. Both system and environment need to be modeled at DT to represent the important characteristics and relations. The introduction of the SDR has increased the flexibility of the wireless terminal and, hence, makes these harder to model. However, as the whole system can be observed in lab environment, it is relatively easy to evaluate and validate system models.

The environment is however more difficult to simulate in a lab environment. Due to many multi-user interactions the environments faced by wireless terminals are

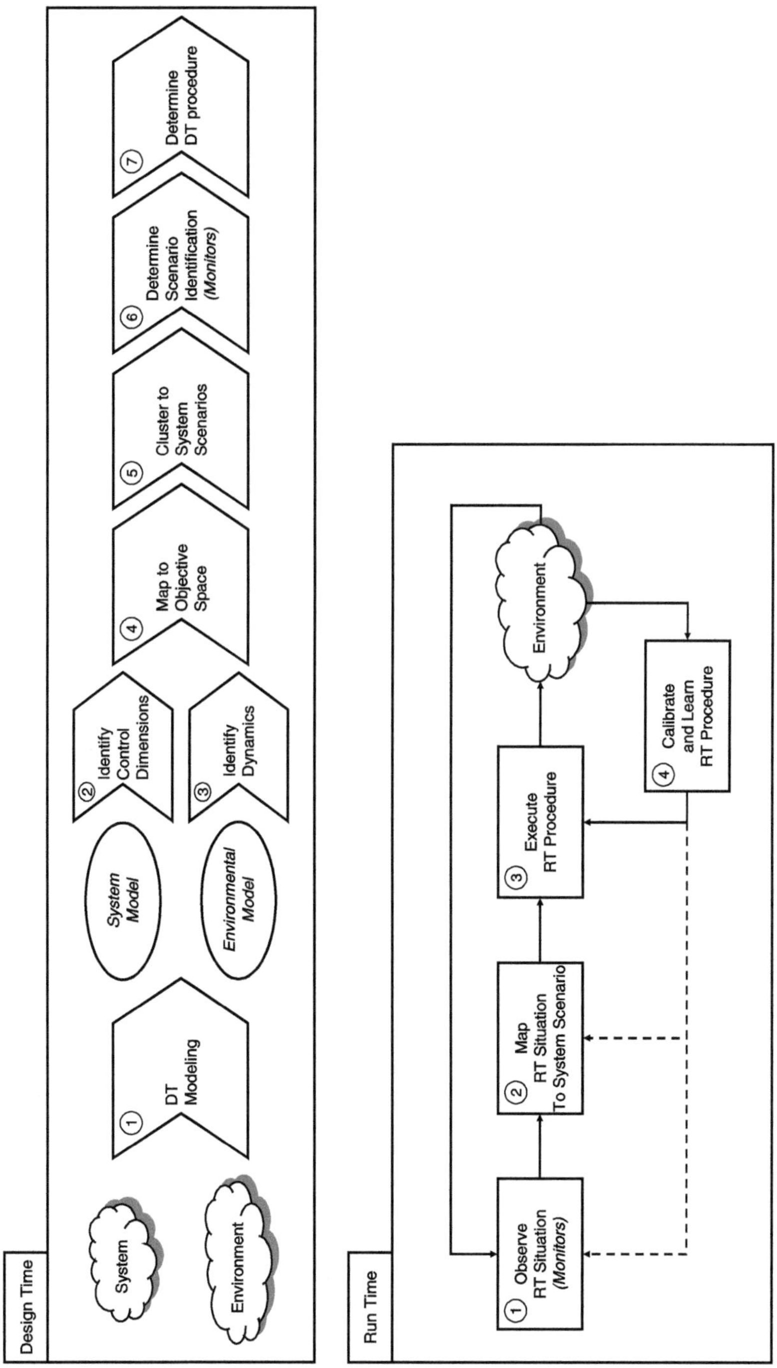

Fig. 3.2 The proposed framework separates DT and RT. While the DT part is a feed-forward process, the RT is a feedback process. This feedback process is essential for CR systems to calibrate and learn procedures at RT

becoming very complex. To make an appropriate DT model, we need to predict at DT, what kind of environments the wireless terminal will face and model these environments. This task is very cumbersome, which is the main reason why validation of research results is evolving from simple channel models over network simulators [42] to elaborate real-life testbeds, e.g., the ORBIT grid of WinLab [43], the IBBT iLab.t [44] or the BOWL project of TU Berlin [45]. The task of validating research results remains very important! However full validation cannot realistically be achieved at DT anymore. As a result, the framework should be flexible enough to adapt to new and unforeseen RT situations. This is heavily supported by the CR paradigm.

3.4.2.2 Identify Control Dimensions②

One of the assumptions of the framework is the availability of control dimensions or knobs to control or steer the system's performance. The focus of this work is on the radio knobs, so the lower communication layers, so no transport and application layer knobs are considered. The number of knobs available in a system, and the range to set them can be implementation dependent. As a control layer to handle a wide variety of control dimensions, the CR is pushing systems to extend their control dimensions.

Control Dimensions (Knobs): $(K_{i,j})$, $1 \leq i \leq n$, $1 \leq j \leq k$. For a given wireless network architecture consisting of n terminals, k control knobs or dimensions exist for each terminal, such as modulation, code rate, output power. The control dimension settings are discrete in real implementations, inter-dependent and together have a complex non-linear influence. We define a setting of all knobs j for terminal i to be configuration point $\mathbf{K}_i$.

In this book, we also consider actions. Actions represent a change over time, from one knob setting to the other. As the system can only control the knobs, it is the only way of acting and interacting with the environment.

Actions: (A_i), $1 \leq i \leq n$. For all n wireless terminals, an action at time t_k is defined as the relative change in its configuration point from t_k to t_{k+1}: $A_i|_{t_k} = (\mathbf{K}_i|_{t_k} \rightarrow \mathbf{K}_i|_{t_{k+1}})$. The actions are selected by the RT procedure and are used to exploit and explore the environment.

3.4.2.3 Identify Dynamics③

As discussed above, the dynamics of the wireless environment are the main drivers to establish an elaborate hybrid DT/RT framework that allows adaptation to these dynamics. The important dynamics are those that influence the performance of the wireless terminal significantly. These should be identified and represented in the environment model.

Dynamics are often difficult to model, especially those resulting from multi-user interactions. In [46], we have illustrated this for large-scale IEEE 802.11 networks. It is hence important to provide procedures flexible enough to adapt their behavior at RT to account for modeling inaccuracies at DT.

3.4.2.4 Mapping to Objective Space④

The identified knobs and environment dynamics can be mapped to an objective space. In this objective space each combination of a certain environment and knob setting corresponds to one point, representing the obtained performance values for all objectives. For best-effort systems, absolute mapping is not necessary. Rather, the actions and environment can be mapped to a marginal objective space which means that actions correspond to an improvement on a certain dimension rather than an absolute performance, which is usually easier to define. An example is for instance a delay constraint: for real-time systems, the delay constraint is a hard constraint, while for best effort systems, the delay constraint is often relative.

3.4.2.5 Cluster and Monitor System Scenarios⑤⑥

In traditional scenario-based design, a scenario describes the use of a future system, in an early phase of the development process. Typically, they appear like narrative descriptions of envisioned usage episodes, called *use-case* scenarios. As in [19, 47], we consider a different and complementary type of scenarios, called system scenarios. These scenarios are used to reduce the system cost by using information about what can happen at RT to make better design decisions at DT and to exploit the time-varying behavior at RT.

System scenarios are the result of a clustering process in the (marginal) objective space and, hence, group RT situations, which have similar performance characteristics (as a function of the knob settings). At RT these system scenarios then need to be able to be monitored and identified. The scenario should be selected that is closest to the current system state. Clearly, a trade-off exists between adaptation accuracy and the number of system states that need to be considered. Each additional system scenario results into memory and monitoring overhead. However, more scenarios result into better adaptation and, hence, increased gains.

System scenario: $(S_{i,m})$, $1 \leq i \leq n$, $1 \leq m \leq s$. As wireless terminals are operating in very dynamic environments, the system behavior varies over time s. Environmental or application dynamics independent of the user or system's control exist, represented by the system scenario variable, $S_{i,m}$.

Monitors: (M_i), $1 \leq i \leq m$. For a given wireless environment, m monitors can be defined. A monitor is an explicit observation of an independent parameter of the observed RT situation. An independent parameter denotes a parameter that the wireless terminal cannot influence. As a result, monitors are a function of the environment and the environment only. This is in contrast to the feedback, which we formally define later in the chapter. Unlike a monitor, the feedback is a function of the environment and the actions taken by the wireless terminal and allows RT calibration.

3.4.2.6 Determine DT Procedure⑦

Finally, a DT procedure is developed. As mentioned before, the DT procedure should be generic enough (parameterizable) to be able to efficiently handle as many RT situations as possible. The RT procedure can be expressed as a fixed procedure that should be carried out. Alternative, the procedure can be defined in a way that enables run time learning or calibration. The RT learning engine then adapts the DT procedure according to the encountered RT situation, which is learned through feedback. A DT procedure is established for each possible scenario. The DT procedure outputs an action distribution, which has to be tuned based on feedback, as a function of the environment.

For instance, in Chap. 7 we introduce the hidden terminal starvation problem. For this scenario, we know that the actions: increase rate, increase carrier sense threshold and decrease power are dominated by their antagonists. The DT procedure for this scenario would then look like:

$$p(\text{decrease transmission rate}) = 0, \tag{3.1}$$

$$p(\text{increase transmission rate}) = \alpha_1, \tag{3.2}$$

$$p(\text{decrease carrier sense threshold}) = \alpha_2, \tag{3.3}$$

$$p(\text{increase carrier sense threshold}) = 0, \tag{3.4}$$

$$p(\text{increase transmit power}) = \alpha_3, \tag{3.5}$$

$$p(\text{decrease transmit power}) = 0, \tag{3.6}$$

where $\alpha_{1\ldots3}$ are the parameters the RT learning engine needs to learn for the observed RT situation.

Alternatively, in Chap. 6 we define a RT procedure for the energy-efficient transmission of multiple terminals in a single wireless network. The RT procedure in that case will be defined completely at DT assuming that the environment, hardware and multi-user interaction can be sufficiently isolated at DT.

Due to switching cost, it can be suboptimal for the terminal to change its behavior after it has detected a scenario change. Hence, possible gains should always be compared with the cost needed to switch from one DT procedure to another.

3.4.3 Run-Time Operation

The challenge at RT is to determine, for each user, the optimal configuration. This requires to monitor the environment, to identify the observed scenario, to execute the RT procedure and to calibrate it. As mentioned in the introduction of this chapter, the different steps in the RT flow are the most important in the smart and cognitive radio operation. As a result, in each of the following chapters, different case studies will be introduced that focus each more on a specific step of the RT flow.

3.4.3.1 Observe the RT Situation①

First, the environment is observed with the monitors, defined at DT. In general, the information to monitor and to be taken into account by the definition process of the smart radio can also be calibrated or partially learned at DT. This is however not considered in this book.

The monitoring of the environment is very important in the context of Opportunistic Spectrum Access, which was probably the first well known example of the cognitive radio paradigm. In this use case, smart radios are called secondary users and have to closely monitor the use of the spectrum by primary users. If a primary user is present in a certain area, the spectrum cannot be used by the secondary user. This simple monitor and act, although the simplest instantiation of the framework introduced in this book, is one of the most known cases of cognitive radio.

3.4.3.2 Map RT Situation to System Scenario②

The encountered scenario is then mapped to a specific system scenario, based on the information derived from the monitors. The observed system scenario gives input to the RT procedure, that is executed and possibly calibrated or even partially learned even at DT. Potentially, the selection of system scenarios can also be calibrated or learned. This is however not considered in this book as we only calibrate the RT procedure (see Fig. 3.2).

A given scenario is often only determined uniquely when monitors across layers and possibly even different nodes are combined. In this book, this will be illustrated for the case of an IEEE 802.15.4 network coexisting with an IEEE 802.11 network. For the IEEE 802.15.4 network to find it's globally optimal channel, it is needed to monitor both the physical layer interference and the routing parameters.

3.4.3.3 Execute RT Procedure③

The RT procedure determines the action to be taken in each of the given scenarios. An action can be an absolute setting of a knob to a value, or a relative increase of the knob setting. For instance, it is possible to give an output power value for the PA, or rather an action can be specified as increasing the output power of the PA with 2 dB.

The actions that should be taken by the smart radio can be fully determined at DT, or only partially specified. In the latter case, the RT procedure is calibrated or learned at RT by means of feedback.

In its most generic version, the RT procedure is hence a function of the current environment, the observed feedback history and the DT procedure.

A RT procedure that is fully characterized at DT is illustrated in Chap. 6. The considered use case is an IEEE 802.11 network where multiple users have to share the medium. By means of scheduling, the power-performance of the entire network is optimized, based on a RT procedure fully specified at DT.

3.4.3.4 The RT Procedure④

When the RT procedure is not fully defined at RT, it is needed to learn and calibrate the actions to be taken. However, to enable learning and calibration, we need to have feedback. As mentioned above, feedback is crucial to develop CR systems. Based on feedback from the environment, CR systems can explore and learn the behavior and response of the environment. Based on this behavior and response the underlying environment models are adapted and new RT procedures can be defined.

Feedback is linguistically defined as [48]:

1. "the process of returning part of the output of a circuit, system, or device to the input, either to oppose the input (negative feedback) or to aid the input (positive feedback)",
2. "a reaction or response to a particular process or activity",
3. "evaluative information derived from such a reaction or response".

Rather than monitors that passively observe the environment, feedback gives information about the environment through its reaction on an action taken by the wireless terminal.

Feedback Channel: (F_i), $1 \leq i \leq f$. For a given wireless environment, f feedback channels can be defined. Rather than passive monitoring, feedback contains information about how the environment reacts to specific actions taken by the wireless terminal. Rather than directly observing the parameters of the RT situations, a feedback channel observes the effect of the actions taken by the wireless terminal on the environment.

As mentioned above, the RT learning engine calibrates the template of the DT procedure. In the example considered in this book in Chap. 7, reinforcement learning is used for the calibration.

The objective of the RT calibration is to find the action that yields the most reward. It is interesting to see that learning algorithms learn a procedure, but do not dictate whether this procedure should be used or not. Exploration beyond the current RT procedure is not only allowed, but advised. Traditionally, this exploration is done blindly. The major problem with blind learning algorithms is the need to explore all possible actions. As a result these algorithms scale badly.

By allowing the DT procedure to steer the exploration of the algorithm away from dominated operating points and by defining possible actions, the number of points that the learning engine needs to traverse can be significantly reduced. Ideally, the DT procedure presents one point, the optimal, which is the case when the RT environment is well known and easily modeled (see Chap. 6). However, as discussed above, for wireless networks this is not feasible anymore.

Reinforcement learning captures the most important aspects of the real problem facing a learning agent interacting with its environment to achieve a goal [49]. Contrary to supervised learning, which learns from examples provided by a genie, and costly techniques, such as neural networks or simulated annealing that rely on the availability of a test set, reinforcement learning learns from interaction with the environment. In interactive problems it is often impractical to obtain examples of desired behavior that are both correct and representative of all the RT situations in which the agent has to act. In uncharted territory, an agent must be able to learn from experience. This makes these reinforcement techniques extremely suited for CR systems.

This case study can be considered as part of a larger trend toward greater contact between artificial intelligence and other engineering disciplines. Obviously, the ultimate CR would be a prime example of such symbiosis.

3.5 Conclusions

In this chapter we have introduced the main challenges in the design of CR systems. It has been shown that those challenges result in a rich set of new design opportunities. We have also given an overview of the system design landscape and motivated our selection for a hybrid DT/RT framework. Afterwards, we introduced our cognitive framework for distributed optimization of best-effort systems. A detailed design flow has then been specified to design systems according to this cognitive framework.

In the remainder of this book, focus will be on the RT steps of the proposed framework. As already mentioned in the introduction, four case studies will be considered that focus on each of the steps: observe RT situation, map RT situation to system scenario, execute RT procedure and finally learn and calibrate the RT procedure.

Chapter 4
Distributed Monitoring for Opportunistic Radios

4.1 To Not Interfere

This chapter focuses on the monitoring challenge that is faced by opportunistic radios. First, the exact problem context of this chapter is introduced. Next, it is illustrated where this fits into the cognitive design flow introduced in Chap. 3. Finally, some measurements are given to emphasize the complexity of this problem.

4.1.1 Problem Context

To solve the spectrum scarcity problem, the reuse of licensed spectrum when it is locally not in use is receiving more and more attention. When such sharing is considered, it is very important that the secondary users adapt their transmission power optimally so that they don't harm the primary user, but simultaneously maximize the throughput of the secondary network. The straightforward approach for this problem is to adapt the transmission power based on local sensing or monitoring or make use of a geolocation database information. The sensing puts significant sensitivity requirements on the sensing device, and moreover requires taking into account large design margins to account for worst-case fading and shadowing of the sensing device (see Fig. 4.1). The database approach relaxes the sensitivity requirements, but is a static approach that is expected to also consider significant margins.

To improve spatial planning of networks, this chapter introduces a distributed and practical monitoring algorithm that

(a) relaxes the sensitivity requirements and hence allows for less expensive hardware and more practical deployments,
(b) achieves optimal power allocation as a function of actual and arbitrary propagation conditions.

Moreover, it will be shown that the algorithm scales well and is robust against measurement noise. Distributed or enhanced monitoring is an important enabler for an

S. Pollin et al., *Software Defined Radios*, Signals and Communication Technology,
DOI 10.1007/978-94-007-1278-2_4,

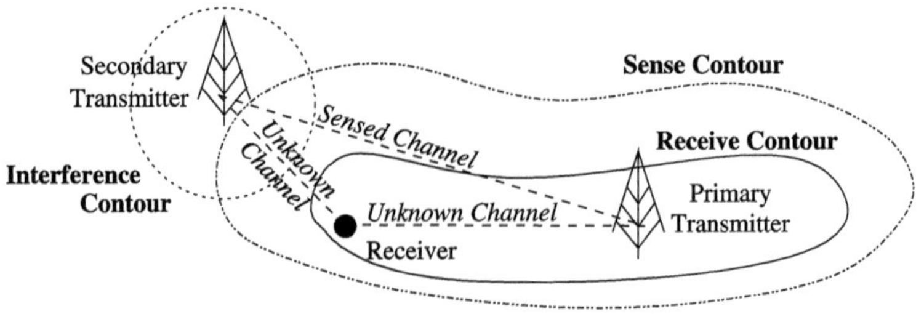

Fig. 4.1 Spatial reuse in wireless networks requires high sensitivity receivers and moreover never achieves optimal adaptation to the real propagation conditions, since safety margins are needed to avoid interference to the potential receivers with unknown channels. We want to achieve optimal spatial reuse (i.e., without safety margins that limit the gain), while relaxing the receiver sensitivity constraints

opportunistic radio. The discussed technique could be used to communicate actual run time propagation contours of (TV) transmitters and improve on the overly conservative power estimates based on design time assumptions.

The challenges of efficient spatial reuse are also illustrated by means of measurements of an outdoor IEEE 802.11 network where one Access Point (AP) is considered to be a primary AP and a second AP should avoid interfering with it.

The technique relies on the availability of nodal positioning information which could be obtained through GPS measurements or a localization algorithm. Future radios are expected to have geolocation capabilities, so this assumption does not put additional constraints on the cognitive radio. When determining the optimal power, a limited amount of interference from control messages is tolerated.

4.1.2 Smart Aspect

The cognitive aspect of the proposed technique lies in the fact that the environment is monitored extensively. The environment in this case is the distance to the primary user receive contour, or more importantly the actual pathloss that is a function of distance but also shadowing. Within that contour, no secondary interference can be tolerated. The secondary network will iteratively adjust it's transmission power until the coverage of the secondary network is within a certain predefined bound of the primary contour. If overlap of the contours is determined, the power will of course be decreased. This means three scenarios are possible: decrease output power, keep output power or increase output power. For each scenario, the corresponding action is then taken until the algorithm converges. This is illustrated in Fig. 4.2.

The approach also relies on a DT model of pathloss and shadowing, that is then calibrated at run-time to determine the actual shadowing, distance and pathloss exponent between the cognitive radio and the primary transmitter. Since explicit feedback is not possible from the primary transmitter (considered to be legacy technology), the information for this calibration is determined based on the monitoring

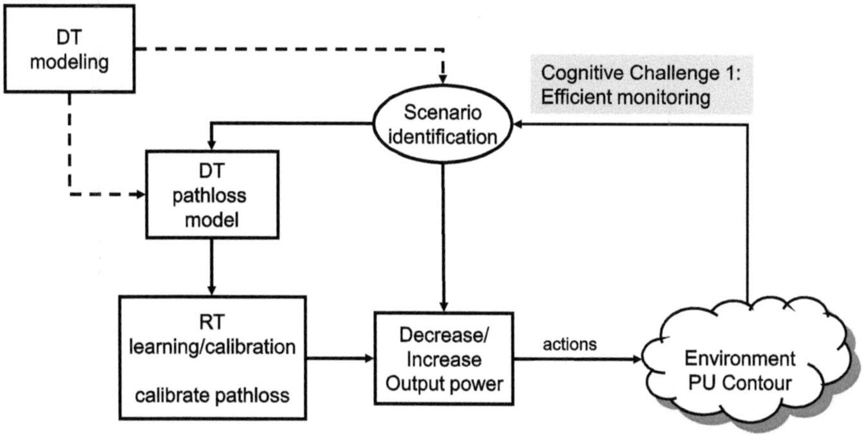

Fig. 4.2 Focus of this chapter is on the monitoring challenge for cognitive radio. The actions are setting the power of the Opportunistic Radio, and these actions are a direct consequence of the monitored environment and a model that translates this environment to a power setting. This model was determined at design time and is not learned by the Opportunistic Radio

information. The algorithm is explained further in this chapter. First, some measurements are discussed to illustrate the complexity of the problem.

4.1.3 Outdoor 802.11 Measurements

When designing the contour estimation and propagation method, it is important to make the correct assumptions. Typically, wireless propagation is characterized by means of a distance-dependent pathloss or trend. On top of this trend, shadowing losses are added that are assumed slowly varying. Finally, fast fading due to multipath effects introduces variations at very small timescales. The impact of each of those components is verified through measurements. This motivates the general contour monitoring method.

4.1.3.1 Measurement Setup

Measurements were obtained from the outdoor 802.11 network at the UC Berkeley campus. A highly accurate GPS receiver (with an error below 1 m) provides location information. Since this is very accurate compared to the pathloss measurements, positioning errors will be neglected. At each location, the Received Signal Strength Indication (RSSI) is obtained which is a measure of received power for each beacon received from the various 802.11 Access Points (AP) on campus.

A reconstructed contour using the algorithm introduced later for one of the APs is shown on the left of Fig. 4.3. The two graphs in the middle and the right of the

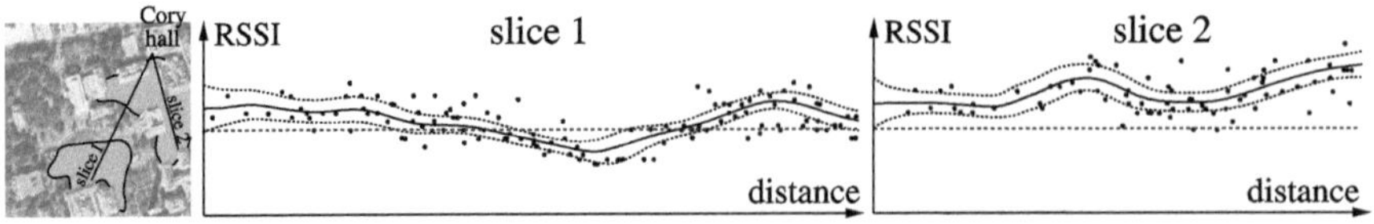

Fig. 4.3 RSSI measurements for an outdoor 802.11 antenna located on top of Cory Hall at UC Berkeley. The *left image* shows the resulting contours for a given RSSI threshold, as we will compute in this chapter. The two *right images* show measurement points and denoised RSSI signals with 95% confidence intervals for two different slices. The horizontal *dashed line* corresponds to the RSSI contour threshold on the *left image*. Note the high noise levels, the absence of a clear trend and the very different results along the two directions

figure show pathloss measurements along the slices indicated on the left overview image. Each of the continuous pathloss curves in Fig. 4.3 is obtained by computing for each point along the horizontal axis a local moving average (details are discussed in Sect. 4.3.2). Along with the moving average the variance is computed, and hence 95% confidence intervals (assuming the error is normally distributed), as indicated in the figures.

4.1.3.2 Observations from Measurements

From these measurements, the following observations can be made.

No Clear Trend: As can be seen in the slices drawn from the measured signal strength in Fig. 4.3, it is not possible to find a clear trend in the pathloss as function of distance. Reception can degrade significantly behind a building, and then improve again at larger distances when the line-of-sight is established again.

Anisotropic Due to Shadowing: As shown in Fig. 4.3, the propagation trend looks very different along the two different paths. It can hence be concluded that it is not possible to fit a single trend to the measurements. Instead we need to fit local trends at run-time, not relying on the assumption that a single model fits the entire propagation surface properly.

Noisy Measurements Due to Fast Fading: The RSSI sample points in Fig. 4.3 appear very noisy. A technique is needed to locally smooth out the noise and obtain the solid curve along with associated confidence intervals.

The main conclusions drawn from the measurements are that:

(1) there is no clear global trend,
(2) propagation conditions vary significantly depending on the direction (since obstacles vary depending on the exact path followed) and
(3) measurements are very noisy.

Clearly, it is very difficult to propose a single pathloss or shadowing model at design time that would fit any run time condition. In this chapter a local channel estimation strategy is proposed to address both (1) and (3). Next, a technique is presented to communicate the local estimate through the network and achieve a power control considering not only distance, but also direction as motivated by conclusion (2). The techniques are based on a design time model that is calibrated at run time based on the monitoring carried out by the cognitive radio.

4.2 The Sensing Problem

Controlling or adapting spectrum use in space is typically achieved using Transmit Power Control (TPC). Many techniques have been proposed to achieve optimal power control through central coordination or in a distributed setting [50]. Each of these techniques however requires a detailed knowledge of the propagation conditions of two conflicting transmitters towards their intended receivers. When it is not possible to rely on such channel state feedback from the receivers, this information is typically obtained through sensing for transmitters. To detect potentially shadowed existing transmitters, the sensitivity requirements of secondary transmitters needs to be set very high [51]. Even when the required sensitivity can be achieved, the allowed power for the secondary transmitter is typically much lower than the power that could actually be tolerated by the primary receiver since large safety margins are required to account for unpredictable shadowing. Indeed, as illustrated in Fig. 4.1, the channel sensed between the primary and secondary transmitter only reflects the channel characteristics along this specific path and safety margins need to be introduced to avoid interference along other paths towards the receivers. An alternative approach is the so-called geolocation database.

4.3 Distributed Distance-to-Contour Estimation

4.3.1 Algorithm Overview and Design Decisions

The problem that should be solved can be stated as follows. For each potential secondary transmitter, this cognitive secondary transmitter should compute its allowable transmission power, so that a maximum number of receiving nodes can be reached without interfering with (potentially multiple) primary transmitters. In the following discussion motivates that this amounts to minimizing received power contour-to-contour distances.

On a high level the algorithm works as follows. First, nodes which sense the primary transmitter compute a robust estimate of the local RSSI. Next, the shortest distance to the interior nodes, i.e., nodes within the propagation contour or with RSSI above a given threshold, is propagated through the network. This distance can be used by a secondary transmitter to conservatively estimate an initial transmission power. Next, the secondary transmitter iteratively estimates the minimal distance between the primary signal threshold contour and its own secondary interference threshold contour and uses this contour-to-contour distance along with a local estimate of the pathloss model to adapt its transmission power to improve the spatial reuse gain.

The design decisions for these three steps which are based on the aforementioned observations in real-world wireless networks are introduced here. The algorithmic details will be discussed in Sects. 4.3.2, 4.3.3 and 4.3.4.

4.3.1.1 Local Channel Estimation

Adapting the spatial resource allocations at the time-scale of fast fading is difficult to achieve in practical scenarios when the time to communicate a change is often much longer than the time during which the change is in effect. The importance of averaging out those fast-fading effects for power control was already noted in [52], where the number of samples needed to achieve a target confidence interval of the slow pathloss estimate is computed for a theoretical fading model.

Since the goal is using power control for spatial resource optimization, it is also required to average out variations over small areas. Large-scale distance-related pathloss and shadowing due to large objects result in large effects that should be considered in the spatial planning. However, fast fading due to multipath effects gives rise to very large variations over distances in the order of a single wavelength. Similar to the motivation above where fast fading effects make it impossible to adapt timely, it is clear that spatial planning using power control cannot take advantage of channel variations over very small areas.

In addition to averaging over time, the secondary transmitter will hence perform (Sect. 4.3.2) a spatial Moving Least Squares (MLS) to obtain a local estimate $\widehat{\mathrm{RSSI}}$ of the received power measure RSSI. MLS is a powerful regression technique to smooth out noise, while obtaining the best local fit of the propagation trend. As mentioned also in [53], and verified by our measurements, it is impossible to rely on a known propagation trend, so fitting this trend adaptively and locally without relying on any false model proves to be a very robust approach.

4.3.1.2 Distance-to-Contour Flooding

Propagating or broadcasting information in a network is typically achieved through flooding. In [54] it is shown that in ad hoc wireless networks, the more important cost factor for flooding is the number of packet forwardings and it is important to prune away unnecessary transmissions.

For reporting a measured value to a sink, many schemes have been proposed in the literature in the context of sensor networks [55]. An interesting approach is found in [56], where the authors study the problem of delivering messages from any sensor to an interested client user along the minimum-cost path in a large sensor network. For a single client, they can establish the optimal cost field with N

messages, through the use of a clever backoff approach. This approach establishes a tree from any sensor to a client. In this chapter a similar algorithm will be used to propagate pathloss information from any receiver to any potential transmitter, hence conceptually resulting in a forest of trees. The resulting flooding method as detailed in Sect. 4.3.3 requires a number of packet forwardings close to the number of nodes N.

4.3.1.3 Iterative Power Control

When each secondary transmitter knows its distance to the first transmitter's contour, it can compute its maximal power as function of the interference margins and a worst-case design time pathloss model as function of distance. From a secondary transmitter's viewpoint, a worst-case pathloss model is a pessimistic one that assumes that the power of the secondary transmitter decays slowly, resulting in maximal interference. Consequently, the computed power level will be much lower than what could be tolerated in reality.

Using the approach of contour estimation and distance flooding, it is however possible to iteratively adapt the distance between the two contours. While estimating the contour of the secondary transmitter, it is also possible to estimate a more realistic pathloss model which in turn can be used to improve the power estimate. Key here is to estimate the pathloss in the direction of the point closest to the first transmitter's contour. Since the pathloss model can vary significantly as a function of the exact path followed, this direction gives the best information for the targeted power control. As will be shown in Sect. 4.3.4, this iterative power adjustment algorithm will converge in only a few steps and will require an order of N communications.

4.3.2 Local Channel Estimation

In this section, it is detailed how a smart secondary transmitter can locally smooth the fast RSSI variations, without relying on a global trend or any prior information, which allows robust inside/outside classification with respect to the threshold contour.

Given a node N_i at location $\mathbf{x}_i$, with received (noisy) power measure RSSI_i, the objective is to find the noise-reduced power $\widehat{\mathrm{RSSI}}_i$. Given a complete polynomial basis $\mathbf{p}(\mathbf{x}) = [1\ \mathbf{x}\ \dots\ \mathbf{x}^n]^T$, the cognitive transmitter tries to find the coefficient vector $\hat{\mathbf{a}}_i$ which minimizes following weighted least squares objective function:

$$\hat{\mathbf{a}}_i = \arg\min_{\mathbf{a}} \sum_j w_{ij} (\mathbf{a}^T \mathbf{p}(\mathbf{x}_j) - \mathrm{RSSI}_j)^2, \tag{4.1}$$

where the summation is over N_i's neighboring nodes N_j (with positions $\mathbf{x}_j$). Note that N_i is contained in its own neighborhood. For example, the complete polynomial basis of order $n = 1$ in two dimensions is $\mathbf{p}(\mathbf{x}) = [1\ x\ y]^T$ and the coefficient vector

$\mathbf{a} = [a_0\ a_1\ a_2]^T$ consists of three unknowns. Results will be given for moving least squares approximations of order $n = 0$, 1 and 2.

The filtered RSSI at the node N_i is then given by:

$$\hat{\text{RSSI}}_i = \hat{\mathbf{a}}_i^T \mathbf{p}(\mathbf{x}_i). \tag{4.2}$$

Locality is obtained by weighting the neighboring nodes' contributions using the distance-based weight function w_{ij} with local support h:

$$w_{ij} = \begin{cases} (1 - r_{ij}^2)^3 & \text{if } r_{ij} \leq 1, \\ 0 & \text{otherwise,} \end{cases} \tag{4.3}$$

where $r_{ij} = (\|\mathbf{x}_i - \mathbf{x}_j\|)/h$.

So, although in practice it is only needed to fit a bi-variate polynomial at each nodal position, the resulting $\hat{\text{RSSI}}$ is defined everywhere and will be C^2-continuous thanks to the above defined smoothly vanishing weight function. This weight function decays fast enough to establish a true local fit [57]. Note that each smart node computes its smoothed $\hat{\text{RSSI}}_i$ estimate independently and only requires local RSSI and GPS localization information to do so, which can be obtained by a single communication step with its neighboring nodes if h is equal or smaller than the communication range of the secondary network users. In the remainder, we will assume that h is the communication range. This step hence requires N communications in the network.

It is easy to see that for order $n = 0$, the minimizer of Eq. 4.1 equals the weighted average measured RSSI:

$$\hat{\text{RSSI}}_i = \frac{\sum_j w_{ij}\text{RSSI}_j}{\sum_j w_{ij}}, \tag{4.4}$$

and thus the moving least squares approximation coincides with the well-known moving average approximation method.

Intuitively, when the main goal of the smoothing operation is to average out spatial variations, the moving average will result in adequate results. Alternatively, in case larger neighborhoods are used and more measurement points are collected, it might make sense to fit a local trend (e.g., linear or quadratic). Such a trend will reduce the smoothing effect on the one hand, but will improve accuracy on the other hand since there are more degrees of freedom for the parameters. This is illustrated in Fig. 4.4, left, using a simulated scenario (see also Sect. 4.3.5) where there is a known contour with a systematically increased noisiness of the RSSI measurements. As can be seen on the figure, when less smoothing is required (less noise power), a quadratic MLS approximates the propagation surface best. However, with increasing noise power more smoothing is required and linear and constant MLS start to outperform.

Choosing the order of approximation as function of the neighborhood size is a delicate trade-off. The choice of local support h as used in Eq. 4.3 influences the approximation quality. As shown in Fig. 4.4, right, an optimal kernel width can be found for all approximation orders and the optimal width typically increases with approximation order.

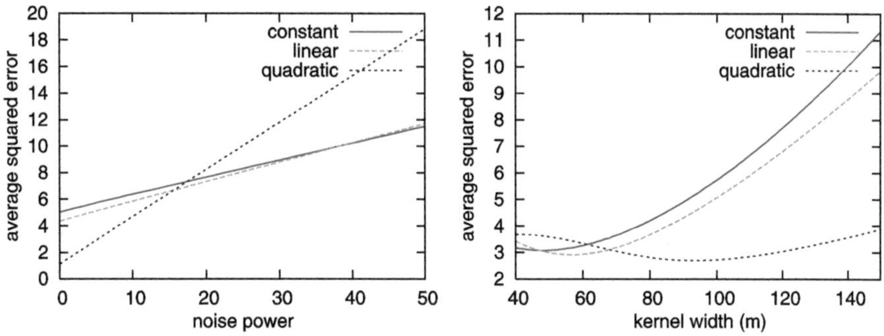

Fig. 4.4 *Left*: Average squared error of the moving least squares approximation (with fixed support radius h) for increasing noise power illustrated for order $n = 0, 1, 2$. At low noise levels, quadratic MLS is superior, while with increasing noise levels, linear and then constant MLS have better performance. *Right*: Average squared error of the moving least squares approximation (with fixed noise power of 4) for varying kernel widths shown for order $n = 0, 1, 2$. Clearly, an optimal communication range h can be found for each approximation order

For the real-world RSSI measurements a noise power of $\sigma_N = 4$ was observed. At this noise level, quadratic MLS with kernel width or communication range of 95 m proved to be the best (cf. Fig. 4.4) and these settings are further used for the simulation results provided in the following sections.

4.3.3 Distance-to-Contour Flooding

A node is now classified as interior with respect to the threshold contour if $\hat{\text{RSSI}}_i \geq \text{RSSI}^{\text{contour}}$. By determining the distance of each node to the closest interior nodes, it is possible to establish an estimate of the shortest distance to the contour. From this distance one can then iteratively approximate the transmit power, as will be explained in Sect. 4.3.4.

A fast marching algorithm will be used which bears similarities to Dijkstra's shortest path algorithm. However, instead of computing a multihop distance, a straight line distance will be computed to the contour, which is possible when GPS coordinates are available. First, the centralized algorithm based on a priority queue is discussed. Next, an approach is presented to make it fully distributed.

4.3.3.1 Centralized Distance-to-Contour Computation

During the execution of the algorithm each node N_i stores its distance d_i to the closest interior node, or footpoint $\mathbf{f}_i$, inside the contour. The algorithm starts by setting $d_i \leftarrow 0$ and $\mathbf{f}_i \leftarrow \mathbf{x}_i$ for all interior nodes. The distance for all other nodes is set to infinity ($d_i \leftarrow \infty$). The nodes with $d_i = 0$ are added to a priority queue, which is sorted in increasing distance to contour order.

The algorithm proceeds iteratively by taking the first node N_i from the queue (i.e., the one with the smallest d_i), and by updating its neighboring nodes N_j (those within a communication distance h to N_i). The footpoint $\mathbf{f}_i$ stored with N_i is used for doing this. If $\|\mathbf{f}_i - \mathbf{x}_j\| < d_j$, i.e., if N_j is closer to N_i's footpoint than it is to its previously computed own footpoint, it is needed to update N_j's distance and footpoint information ($d_j \leftarrow \|\mathbf{f}_i - \mathbf{x}_j\|$ and $\mathbf{f}_j \leftarrow \mathbf{f}_i$) and add N_j to the priority queue. This process is repeated until the queue becomes empty.

Due to the use of a priority queue this algorithm requires a central base station. However, it is possible to approximate this algorithm and make it fully distributed by cleverly modifying the medium access protocol as will be discussed below.

4.3.3.2 Distributed Distance-to-Contour Flooding

When implementing the fast marching in a network, it is required that each node updates its neighbors by sending its footpoint information. We note that nodes that are inside the contour are footpoint to themselves, and hence have perfect information at the start of the algorithm. Inspired by the optimal centralized implementation of the algorithm, one can see that it is optimal to have nodes with a smaller distance to the contour update their neighbors first. In practical networks, this can be achieved by giving those nodes a higher probability p_{send} to gain access to the channel. While nodes are waiting to access the channel, they can accumulate neighbor update events and only propagate the best footpoint found so far. This probability p_{send} is typically a function of a backoff timer in practical protocols such as 802.11, or alternatively, it is a function of the scheduling algorithm carried out at the AP.

The simulations are done with a random backoff timer that determines when nodes update their neighbors. Assume that CW is the initial backoff range that results in a low number of packet collisions that can be neglected. A backoff timer can be used to determine when nodes can access the channel, by choosing a random number in the backoff range. Since distance-to-contour information is spreading in the network away from the contour, one wants to make sure that nodes only send when all nodes in their neighborhood that are closer to the contour have sent first.

As illustrated in Fig. 4.5, one can be sure that all black footpoints f_i with backoff in the range $[0 \ldots CW[$ will have sent after time CW, if collisions requiring retransmissions are neglected for the ease of reasoning. As a result, a node A that receives an update from a footpoint f_1 (with backoff B_{f_1} in range $[0 \ldots CW[$) will have to wait an additional $CW - B_{f_1}$ after reception of f_1's update. This can be achieved if A will select a random number in the range $[(CW - B_{f_1}) \ldots (CW - B_{f_1}) + CW[$. Clearly, as indicated in Fig. 4.5, the updating speed of the network will be approximately[1] the one-hop communication range h, each CW timeslots.

A distributed implementation of the scheme potentially results in a node spreading information before it obtained its closest footpoint. Such a node will have to

[1]Due to sampling irregularities, as illustrated in Fig. 4.5 also, the contours do not spread exactly over a distance h each step.

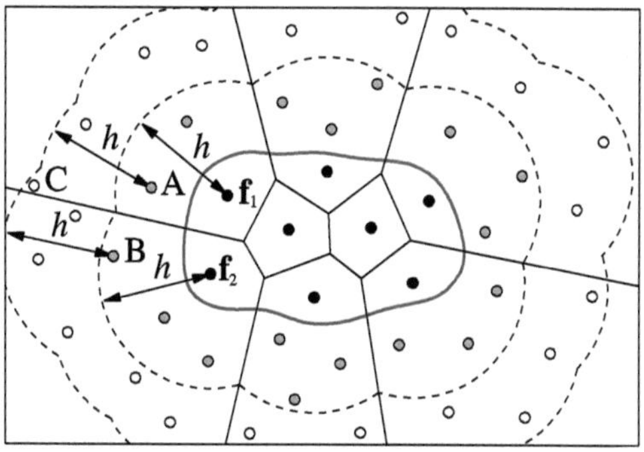

Fig. 4.5 *Black nodes* (such as f_1 and f_2) correspond to interior nodes (i.e., nodes inside the contour). The *straight lines* trace the regions which are closest to a certain interior node (i.e., the *Voronoi* regions). The new proposed backoff scheme ensures that nodes (such as A and B) which are closer to the interior contour nodes are updated first before they propagate this contour distance information to the other nodes. This propagation can be performed at a cost of almost N transmissions, the total number of nodes

update its neighbors more than once. This is illustrated for node C in Fig. 4.5. Node C receives information of node B only before time $2CW$, and hence thinks its closest point on the contour is f_2. Following the algorithm, C will have chosen a random backoff in the range $[(CW - B_B) \ldots (CW - B_B) + CW[$ to start updating its neighbors. If node C does not receive correct information before that random time, wrong information will be propagated and node C will have to update his neighbors a second time once information about the correct footpoint f_1 is obtained. In practice, with this scheme, the number of update events is experienced to be very close to the number of nodes (neglecting collisions) as also illustrated in Table 4.1. The contour spreading as function of time is illustrated in Fig. 4.6 for a simulated scenario.

It is important to see that the algorithm will converge for arbitrary complex contour shapes, such as the one shown in Fig. 4.3 which consists of multiple parts.

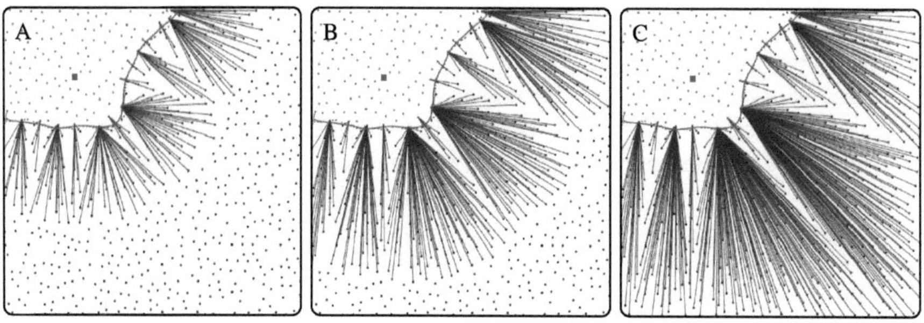

Fig. 4.6 Distributed distance-to-contour flooding shown at three intermediate timesteps during the algorithm for a simulated scenario

4.3.4 Iterative Power Control

Given the learned shortest distance to the target contour, the cognitive secondary user decides conservatively on its transmission power. This initial transmission power is typically suboptimal and will now be adapted efficiently by the following iterative procedure. Using the contour estimation method proposed in Sect. 4.3.2, nodes classify themselves as interior or exterior with respect to the secondary sender's target interference power contour. If no node inside this contour is within a target distance of the first transmitter's contour, the secondary transmitter should increase its power. The amount of power increase will be robustly determined by locally estimating the pathloss parameters as will be discussed below. Alternatively, in case the contours overlap, the cognitive sender's power will have to be decreased. The number of messages needed to inform the secondary transmitter will be the number of interior nodes N_{interior}, which is smaller than N.[2] This power adaptation is iterated until the distance between the contours is smaller than the one-hop communication distance h. Both scenarios (i.e., power increase and power decrease) are now discussed in detail.

4.3.4.1 The Increasing Power Scenario

While estimating the second sender's propagation contour, interior nodes also robustly approximate the local pathloss parameters $\hat{\alpha}_i$ and $\hat{\beta}_i$ by minimizing following weighted least squares objective function:

$$\{\hat{\alpha}_i, \hat{\beta}_i\} = \arg\min_{\alpha,\beta} \sum_j w_{ij}(\alpha 10\log_{10}(r_j) + \beta - \text{RSSI}_j)^2, \tag{4.5}$$

where $r_j = \|\mathbf{x}_j - \mathbf{x}_{s2}\|$ is the distance from node N_j to the secondary sender and α is the pathloss exponent and β represents system losses.

The summation is performed over the nodes N_j in the local neighborhood around N_i, and the weighting is distance-based as given by Eq. 4.3.

Due to this locality, each interior node N_i will find a different estimate of those parameters. However, as mentioned earlier in Fig. 4.3, it is indeed the case that pathloss can vary significantly as function of the exact direction. In the power control problem, it will be most useful to use the pathloss estimate of the node that will be closest to the first transmitter's contour.

Interior nodes N_i now send a message to the sender which includes following information: estimated pathloss parameters $\hat{\alpha}_i$ and $\hat{\beta}_i$, footpoint $\mathbf{f}_i$ on the primary sender's contour and distance to this footpoint $d_i = \|\mathbf{x}_i - \mathbf{f}_i\|$. The cognitive secondary sender retains from all these messages the one with minimal d_i. The corresponding node N_i is the interior node (with respect to the secondary sender's

[2] Here, it is assumed that each interior node can reach the secondary transmitter in a single hop, which is the case in most communication systems that are duplex.

contour) which is closest to the primary contour and hence considered to give the best estimate of α and β.

The distance $d = \|\mathbf{x}_{s2} - \mathbf{f}_i\|$ from the secondary sender $\mathbf{x}_{s2}$ to the footpoint $\mathbf{f}_i$ is used together with the estimated pathloss parameters to estimate an improved transmission power:

$$P = P_{\text{interference}} + \hat{\alpha}_i 10 \log_{10}(d) + \hat{\beta}_i, \tag{4.6}$$

where $P_{\text{interference}}$ denotes the maximal interference that can still be tolerated at the established contour of the first sender.

This process of power adaptation based on the estimated pathloss parameters at the minimal contour-to-contour node can be iterated to further improve the secondary transmission power estimation. As can be seen in Fig. 4.7, this requires only a few iteration steps. In each step, one contour estimation and information propagation substep needs to be carried out, each with complexity of at most N messages.

4.3.4.2 The Decreasing Power Scenario

When the contours overlap, it will be required that the opportunistic transmitter decreases his transmission power. This however is relatively easy, since a node i inside both contours will have an estimate of the received power from both transmitters, and can hence easily determine the ΔP_i causing the overlap in contours. A node noticing such an overlap simply communicates the ΔP_i to the second transmitter, that follows the largest power adjustment reported. This case is very unlikely to happen, but since we extrapolate an estimated pathloss model, we cannot avoid the possibility.

4.3.5 Results

In this section, the simulation setup is introduced and some results discussed.

4.3.5.1 Simulation Model

In order to verify the approach on a data set that is not subject to measurement errors, it is appropriate to set up a simulation model. The simulation scene consists of a two-dimensional area in which buildings or obstacles result in shadowing losses L_S, on top of a general pathloss trend that varies with distance:

$$L = 10\alpha \log_{10}(d) + \beta + L_S + X_\sigma, \tag{4.7}$$

where L is the total pathloss in dB ($\alpha = 4$ and $\beta = 40$ dB for the simulations). Since the measurement data shows a lot of variations in the measured RSSI, even for a fixed position d and shadowing L_S, zero-mean Gaussian distributed variations or

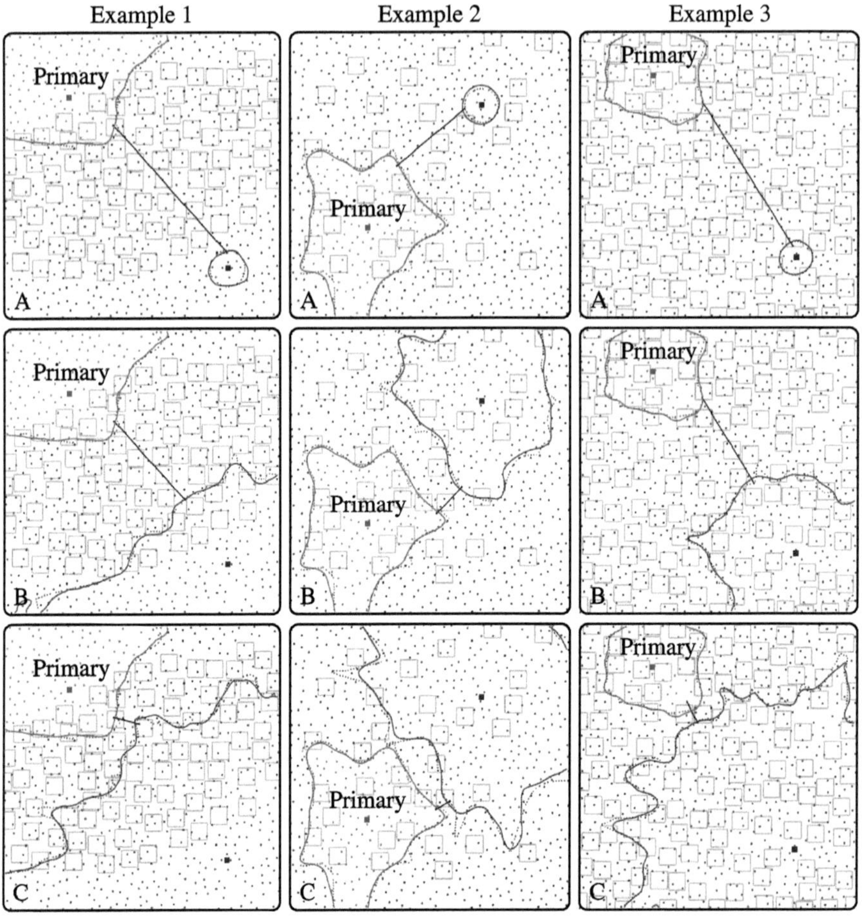

Fig. 4.7 Iterative power adjustment for 3 simulation scenarios of 1 km^2 with communication range $h = 95$ m, quadratic MLS and $\sigma_N = 4$ which is the variance noticed on the outdoor measurements. Each *box* represents a building causing shadowing losses. The *solid curves* represent the approximation of the primary signal's and the secondary transmitter's interference contour. The *dotted lines* are the real (noise-free) contours. The *straight line* on each image corresponds to the contour-to-contour distance as computed by our algorithm. Row **A** shows for each example the resulting contour for the initial suboptimal power. Note the small area covered by the initial secondary contour. From the estimated local pathloss model at the secondary transmitter's contour point, a new transmission power is computed. Rows **B** and **C** show the result of this iterative process. Note the large difference between the secondary transmitter's coverage area at the original estimate in **A** and the resulting contour in **C**

noise X_σ is added. Following the ITU Terrain Model [58] L_S is modeled as function of the depth of the line-of-sight ray into the building. The motivation behind this is that the wavefront can be modeled as a zone of energy around the line-of-sight. Diffraction losses occur when part of this zone is obstructed. When the ray just passes next to the building, half of the zone is obstructed. When the ray passes in

Table 4.1 Performance of the algorithm for the examples of Fig. 4.7

		Number of updates	Percentage misclassified
Example 1 (710 nodes)	primary sender	724	0.28
	secondary sender	721	0.14
	secondary sender update 1	722	0.98
	secondary sender update 2	714	1.12
Example 2 (723 nodes)	primary sender	732	0.83
	secondary sender	730	0.14
	secondary sender update 1	732	1.24
	secondary sender update 2	732	2.21
Example 3 (712 nodes)	primary sender	721	0.70
	secondary sender	723	0
	secondary sender update 1	722	0.14
	secondary sender update 2	719	0.98

the middle of the building, the whole zone is obstructed, leading to larger losses. In this simulation study, the model is slightly adapted to allow received signals inside buildings. The exact form of the model is not that important since our technique should work for any possible situation. The resulting shadowing loss is

$$L_S = \max\left\{0, \left[5.125 \times \left(2 - \frac{\|\mathbf{x}_L - \mathbf{x}_O\|}{D/2}\right)\right]\right\}, \tag{4.8}$$

where $\mathbf{x}_O$ is the position of the obstacle, D its width and $\mathbf{x}_L$ the orthogonal projection of $\mathbf{x}_O$ on the ray. The fraction $\frac{\|\mathbf{x}_L - \mathbf{x}_O\|}{D/2}$ tells us how close to the center of the building the line-of-sight ray crosses. When the line-of-sight ray crosses the building through the center, we have a maximal L_S of 10.25 dB per obstacle which is a value chosen from [51]. From the center of the building, the shadowing loss decreases linearly until 0 dB.

4.3.5.2 Results and Discussion

The resulting iterative power adjustment is illustrated for 3 examples in each column of Fig. 4.7. The primary sender's contour approximation is drawn using a solid line. The real contour is drawn as the dashed line. The computed minimal contour-to-contour distance is indicated by the straight line connecting the two corresponding nodes. The buildings are drawn as gray boxes. The simulation model as described above was used, with a noise power of 4 corresponding to the real-world measurements. We used quadratic MLS with a kernel width of 95 m in all examples.

As can be seen in all images, the computed contours approximate the exact ones very well: Only 0.7% on average of all nodes was misclassified. Note that these

2D contour curves are in reality not constructed, but nodes locally make a binary classification as being interior or not. The curves are shown only for illustration purposes.

The images in row A show the secondary contour resulting from the initial power estimate based on the distance between the secondary sender and the primary contour. An optimistic pathloss model with $\alpha = 2$ was used for the computation of the secondary transmitter's power as function of the measured distance from the first transmitter. Since actual propagation is by definition worse than the most optimistic condition, the coverage of the secondary transmitter is typically very small using this power, as can be seen in Fig. 4.7, row A.

The algorithm however iteratively estimates a more realistic pathloss model for the secondary transmitter, and as a result achieves a much more realistic power adjustment already after one iteration. As can be seen in Fig. 4.7, the resulting coverage of the secondary transmitter can be increased considerably in each of the 3 scenarios, without causing interference (i.e., overlap of the contours) to the first transmitter. Each power adjustment step requires estimation of the secondary transmitter's contour at the cost of N transmissions. Next, each node inside the contour requires propagating its distance to the first contour and its estimated pathloss, at the cost of $N_{\text{interior}} < N$ transmissions. As can be seen in the final images C, a much larger area is covered by the secondary transmitter as compared to the initial configuration shown in row A.

The algorithm not only results in improved spatial reuse gain, but also relaxes the opportunistic transmitters' sensitivity requirements. Indeed, traditional approaches would require each possible transmitter to be able to sense the first transmitter. For a given target power of the second transmitter, this results in very high sensitivity requirements. In this distributed run time approach, these requirements are relaxed, and the first transmitter's propagation contour is estimated close to the first transmitter and then efficiently propagated. Both this estimation and propagation are cheap and require only N transmissions. Typically, the algorithm converges in two to three iterations.

Due to the approximative inside/outside classification, we can not guarantee that the real final contours do not overlap and little interference might occur. As can be seen on the images, only in example 2 the final contours overlap, though for a negligible amount. However, it would be interesting future research to quantify the amount of possible overlap.

4.4 Conclusions

In this chapter, focus was on the monitoring challenge for smart opportunistic radios. A new approach for iterative power control where to goal is to maximize spatial reuse while avoiding unwanted interference to existing networks was proposed. First, the propagation contours of the existing networks are estimated and efficiently propagated. Next, the secondary transmitter's power control is iteratively adapted based on estimates of real pathloss conditions. This is done by calibrating a design time model by means of run time monitored information. No explicit feedback from the primary system can be assumed in this case. The main concepts behind the design of the scheme were motivated from outdoor measurements. The resulting scheme is shown to result in a much improved spatial reuse compared to traditional approaches. By instantiating the framework, leveraging on concepts such as design time modeling and run time calibration and monitoring, it is possible to avoid interference while optimizing spatial reuse.

Chapter 5
Coexistence: The Whole Is Greater than the Sum of Its Parts

5.1 Introduction

We have argued in Chap. 3 that RT solutions based purely on DT models are ineffective. In a complex environment solutions need to be learned at RT. In this chapter, we build further on this observation. We focus on the horizontal sharing of two standards that operate in the 2.4 GHz ISM band: IEEE 802.11g Wireless LAN [59] and IEEE 802.15.4 Sensor Networks [60], illustrated in Fig. 5.1. As mentioned in the previous chapter, in the ISM bands wireless spectrum sharing is the most flexible and, hence, the most difficult to control. Though the IEEE 802.11 is the most popular standard and intra-coexistence is important (also see Chap. 7), the ISM bands are horizontally shared by many other devices like IEEE 802.15.4 sensor terminals. From Fig. 5.1 we can see that, while these two standards have equal rights to the spectrum, the characteristics of both standards are very different and the problem turns out to be asymmetric in nature.

Indeed, the output power of IEEE 802.15.4 devices is usually as low as 0 dBm [61], while the output power of IEEE 802.11g devices is typically above 15 dBm [62]. Also, IEEE 802.15.4 sensor networks are designed to monitor the environment or buildings, and are typically very large. IEEE 802.11 networks are typically hotspots centered around an Access Point (AP), and hence more local. Finally, sensor network applications are not demanding in terms of throughput, but however require a high reliability and robustness against attacks or unknown events. They should also be self-organizing since it is impossible to maintain such large networks manually. IEEE 802.11 networks are typically used by a limited number of throughput-intensive applications.

Although recent studies have shown that sensor networks do have an impact on IEEE 802.11 networks [63], it is well accepted that coexistence of both networks affects the sensor networks most [64]. In this chapter, we design distributed protocols for IEEE 802.15.4 networks to optimize their performance under varying interference. The proposed algorithms should be distributed in order to improve scalability, robustness and adaptability. More specifically, we design distributed channel selection algorithms that allow the sensor terminals to dynamically adapt their channel

S. Pollin et al., *Software Defined Radios*, Signals and Communication Technology,
DOI 10.1007/978-94-007-1278-2_5,

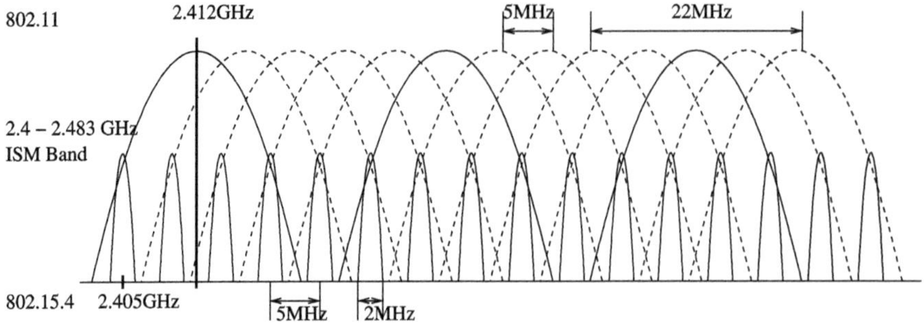

Fig. 5.1 IEEE 802.11 and IEEE 802.15.4 both operate in the 2.4 GHz ISM band. This leads to coexistence issues, most prominently at the side of the ZigBee network

selection to the observed interference patterns. Importantly, the algorithms are evaluated for coexistence with IEEE 802.11 systems. However, they are designed to be so flexible that they can adapt to any interference pattern, as long as the coherence time of the interference is longer than one IEEE 802.15.4 period.

This chapter is structured as follows. First, we explain the system model in Sect. 5.2. The benchmark solution for this chapter is presented in Sect. 5.3. In Sect. 5.4, we describe the problem statement and how we can reformulate the problem for practical use. We also detail the downsides to the problem reformulation. In Sect. 5.6, we design learning-based algorithms, that eliminate the scanning overhead. In Sect. 5.7, we show that these energy-efficient learning-based algorithms outperform the scanning-based approaches. Finally, in Sect. 5.8, we present out concluding remarks.

5.2 Modeling Coexistence

In this section we give a short overview of the models used for the sensor network, the WLAN interference, and the considered energy and performance metrics that are relevant for the considered scenario. For a more detailed overview of the assumptions made and the implementation details, we refer to [19, 65].

5.2.1 IEEE 802.15.4 Network Model

Sensor networks typically consist of a large number of terminals, e.g., to monitor the environment or for building automation. As a result, we represent the IEEE 802.15.4 network by a large number of terminals n that are arranged in a string. The connectivity matrix $\mathbf{C}_{(n,n)}$ denotes which sensors can overhear which others. We assume that all sensors in the range Rd can overhear each other's beacons, where d is the inter-terminal distance. In Fig. 5.2 a simple string topology is presented with

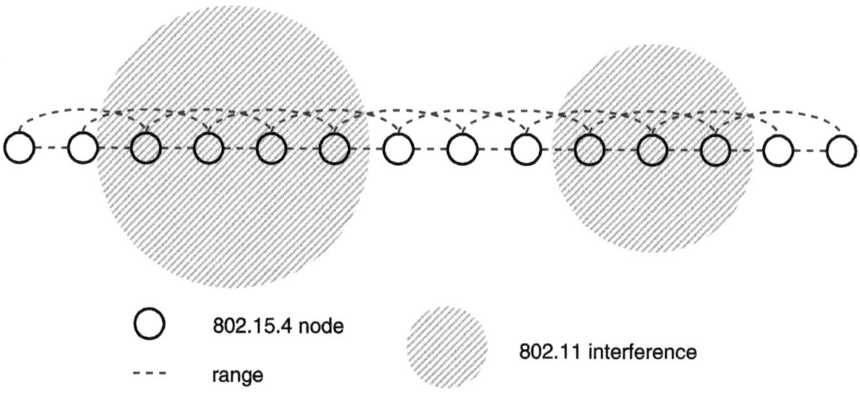

Fig. 5.2 The considered scenario is a string topology of IEEE 802.15.4 terminals, where terminals report to a sink that is placed at one side of the string. Interference is generated by WLAN devices and is dynamic in both time and space

$R = 2$. Each of the sensor terminals is currently operating in one of the F possible channels. As explained above, $F = 16$ for IEEE 802.15.4 networks operating in the 2.4 GHz ISM band. An n-dimensional vector $\mathbf{t}_n$ keeps track of the current frequency f_i $(i \in [1, \ldots, F])$, which the different terminals are using to transmit. Every terminal transmits its beacon at this frequency, according to the beacon-enabled mode. We assume that the sensors potentially swap frequency every inter-beacon period, which ranges from 15 ms to above 4 min according to the standard. When and how to swap will be determined by the distributed adaptation algorithms.

5.2.2 IEEE 802.11 Interference Model

We consider a large IEEE 802.15.4 network, that is affected by IEEE 802.11 interference. This IEEE 802.11 interference is assumed to vary over time, space and frequency.

The variations in frequency result from the fact that different IEEE 802.11 networks can operate simultaneously using a different channel. Dynamic Frequency Selection (DFS) is a new functionality currently added to most of those IEEE 802.11 networks and Access Points. It is developed to optimize the frequency allocation of IEEE 802.11 networks that are subject to interference, which can be self-interference or interference resulting from other networks or devices. As a result of this adaptive and hence dynamic frequency selection, IEEE 802.11 networks are not operating at fixed channels over time.

It can be seen in Fig. 5.1, that each IEEE 802.11 network always covers four consecutive channels of IEEE 802.15.4 networks. The power distribution over the channel is more or less flat, especially for the OFDM-based IEEE 802.11g networks [59].

The variation in time results from the fact that the activity of the IEEE 802.11 networks is dependent on the time-varying demand of its users [66]. From this study it can also be reasonably assumed that the IEEE 802.11 interference stays constant over several (thousands) of IEEE 802.15.4 inter-beacon periods.

The spatial variations result from the IEEE 802.15.4 deployment, which usually covers a very large area (e.g., for monitoring purposes). As a result, the IEEE 802.15.4 networks are expected to be large both in terms of the area they cover and the number of terminals they consist of. The theoretical transmission, and hence also interference, range of IEEE 802.11 networks is 100 m to even 250 m. Although this is a significant range, sensor networks can cover larger areas since they consist of a large number of terminals in a mesh topology. As a result, the IEEE 802.11 interference is assumed to affect large geographical subsets of the IEEE 802.15.4 network of terminals (Fig. 5.2).

As a last step, we assume that an active or interfering IEEE 802.11 network will always be detected by the IEEE 802.15.4 terminals [65].

The IEEE 802.11 interference can thus finally be modeled as a matrix $\mathbf{I}_{(n,F)}$. Each interfering network then corresponds to a submatrix of dimensions $(n_i, 4)$, where n_i denotes the number of terminals that are interfered by network i (depending on its output power), and where it is taken into account that every IEEE 802.11 interference pattern has a width of four IEEE 802.15.4 channels. Networks can swap frequency over time, disappear or appear, but this time variation is assumed to be slow compared to the IEEE 802.15.4 frequency adaptation.

5.2.3 Performance and Energy Measures

We consider delay as a relevant performance metric (throughput requirements in sensor networks are typically low). More precisely, assuming that sensors monitor a variable that should be communicated to a central sink, we consider the number of periods required to forward a measurement to a fixed central sink. This average is computed over time and over the terminals in the network. The more the network is affected by the interference, the more periods are required on average to reach the sink. We assume that every packet is forwarded during each period, to the terminal closest to the sink that can be reached in that period. As a result, packets travel each period the largest possible distance.

For the energy cost, we can model the energy needed during every period independently of the actual packets sent, received or beacons overheard. This is a valid assumption as throughput in sensor network applications is typically very low, and, moreover, the transmit power, P_{Tx}, is lower than the receive power, P_{Rx} [30]. Since the full receive chain is required to be on for scanning, the power consumption in that mode is the same as the power cost in the receive mode! During every superframe, each terminal is awake to listen at least to its current frequency channel. The quality of a frequency channel can be assessed by counting the overheard beacons of neighbors. If no beacons are heard, energy detection, which is part of the

IEEE 802.15.4 specifications, can be used to detect interference on the channel. As a result, the energy consumption can be modeled as only varying with the number of channels that are scanned (or listened to) in parallel:

$$E_{\mathrm{tot}} = n_{\mathrm{s}} t_{\mathrm{a}} P_{\mathrm{Rx}}, \tag{5.1}$$

where n_{s} is the number of channels scanned and t_{a} is the active period per superframe.

In sensor networks, switching energy also presents a large portion of the consumed energy. As we explain in Sect. 5.4, we redefine the problem statement. The goal of the current chapter is to have all terminals on the same channel, so switching isn't necessary anymore. This way, a terminal that performs well in the redefined problem, also minimizes switching energy.

5.3 Basic Solution: Random Frequency Selection

The simplest DFS algorithm is a scheme where nodes randomly (uniform distribution) pick a channel every period. It can of course be expected that the average delay in this Random Frequency Selection (RFS) scheme will be large. However, since it does not rely on any coordination between the nodes and does not rely on an environment model, it is very robust.

5.4 The Problem from a Different Angle

In the current chapter, our goal is to minimize the average packet delay to the sink. This target function is unfortunately quite difficult to model, so we will approximate it by inspecting the optimal solution. In any learning system, defining a target function, which can be measured at RT, is a vital step in the solution design (see [67, Chap. I]).

As mentioned in Sect. 5.2, we assumed the sensor network to have a low duty cycle. Under this assumption a packet has the least delay if it can travel the furthest distance each hop. Obviously, the optimal solution is then the one where all terminals share a common channel. In that case, we can ensure that each packet, independently of where it was generated, can travel the furthest distance and have an efficient last hop.

We can now come up with a simple approximation of our target function: we no longer look at the delay, but try to get all the terminals on the same channel. Clearly, any solution presented in this chapter can thus also be used to dynamically and distributed find a common control channel (CCC), which is of high importance for Opportunistic Radio networks [68, 69].

5.5 Scan-Based Approaches

In [19, 65, 70], we have shown that it is possible to outperform RFS, since the considered IEEE 802.11 interference does not vary every IEEE 802.15.4 period. Here, we explain how scan-based approaches relying on techniques similar to simulated annealing, can be elegantly implemented in the considered network setup.

In Sect. 7.3.5, we use simulated annealing can be applied to the distributed optimization of IEEE 802.11 networks. In this section, we use a similar technique, that only considers the potential energy as a variable. Hence, we don't anneal. In Sect. 5.6.3.2, we explain in detail why it is very difficult to anneal in this context.

When optimizing the frequency allocation over a large sensor network affected by dynamic interference, terminals need to keep looking for another channel. Every period, next to its current frequency channel, f_i, terminal i scans n_s other channels (grouped in $\mathcal{S}_i$) and assesses their performance. This is done according to a reward function R_w.

We use the following definition of $R_w(f)$ as an approximation of the real target function (also see Sect. 5.4):

$$R_w(f) = \begin{cases} \sum(\text{beacons heard}) + 1 & \text{when no interference detected,} \\ 0 & \text{when interference is detected.} \end{cases} \tag{5.2}$$

When IEEE 802.11 interference is present, the reward for using that channel is equal to zero (worst-case). When no IEEE 802.11 interference is present, the reward function is assumed to be proportional to the number of heard beacons, augmented by one to distinguish from the worst-case.

In simulated annealing, exploration is embedded in the algorithm to allow the system to jump out of a local optimum. This means that a scanned channel can be accepted even if it is measured to be worse than the current channel according to the observed reward. Usually, this happens with a certain probability that should be decreased each time interval to eventually converge to the optimal solution.

The group of candidates for possible selection, $\mathcal{C}_i$ is defined as follows:

$$\mathcal{C}_i = (f_i \cup \mathcal{S}_i) \backslash \mathcal{I}_i, \tag{5.3}$$

where f_i is the current channel, $\mathcal{S}_i$ is the group of scanned channels and $\mathcal{I}_i$ are the channels where terminal i detected interference. The probability to select a channel, f, from the group of candidates is then equal to:

$$p(f) = \frac{e^{\frac{R_w(f)}{T}}}{\sum_{\forall f \in \mathcal{C}_i} e^{\frac{R_w(f)}{T}}}, \quad f \in \mathcal{C}_i, \tag{5.4}$$

where T is a constant temperature.

As far as energy is concerned, the proposed algorithm requires to scan the current channel and n_s. Scanning cost is, hence, multiplied with a factor n_s as compared to RFS, where only the current channel is scanned.

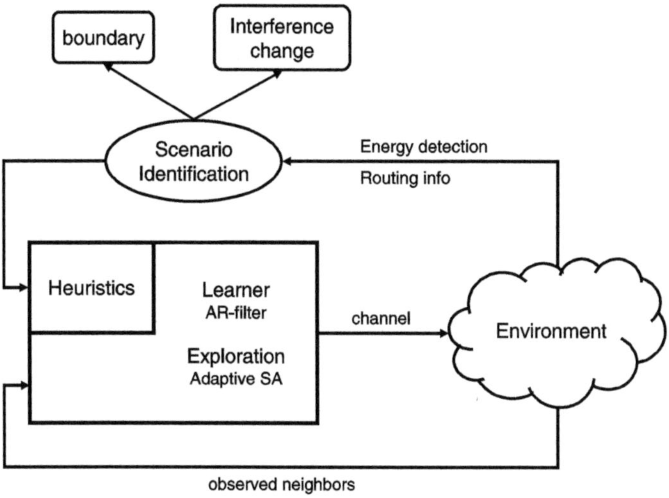

Fig. 5.3 The framework of the dynamic frequency selection algorithms relies on feedback from the environment. The performance increase of the learning engine allows to decrease the quality of the feedback (no out-of-band scanning), while maintaining similar performance (see Sect. 5.7)

5.6 Distributed Learning and Exploration

In this section, we explain how we can design scanless approaches, using the generic framework, presented in Chap. 3. The goal of these scanless approaches is to reach similar or better performance as those presented in Sect. 5.5, while removing the extra hardware complexity and energy cost. Rather than extensively monitor the scenario, which requires scanning overhead, in this section we *learn* at RT. As mentioned in Chap. 3, this is a vital element of Cognitive Spectrum Access.

5.6.1 General Framework

In this section, we give an overview of the proposed algorithm and how it relates to the generic framework, presented in Chap. 3. In Fig. 5.3 the framework of the entire control algorithm is shown. The DT procedure used is simple RFS, presented in Sect. 5.3. From this DT procedure, a RT procedure is learned through interaction with the environment. The framework relies heavily on the feedback from the environment. The number of observed neighbors is used to learn an optimal configuration. In this case study chapter, we have selected a simple AR-filter as learning engine. This is further detailed in Sect. 5.6.2.

The exploration algorithm is discussed in Sect. 5.6.3. We have already mentioned earlier that it is very difficult to anneal in the current context. This is further explained in Sect. 5.6.3.2. In Sect. 5.6.3.3, we explain how a scenario-based approach can overcome the obstacles faced when annealing. The considered scenarios are boundary detection and appearance of IEEE 802.11 interference. For both scenarios, the heuristic recommendations increase the exploration rate of the engine.

5.6.2 Learning Engine

In this chapter, we consider a simple Auto Regressive (AR)-filter. Contrary to Q-learning (see Chap. 7 and [67]), the AR-filter maintains a one-dimensional table rather than a two-dimensional one. This has no impact on the performance in the current context as the reward function is independent of the current state of the terminal.

Our goal in this section is to learn the reward of selecting a certain channel. To do so, we estimate the reward function, $R_w(f)$, presented in Sect. 5.5. This is done by updating its current estimate using the following rule:

$$\widehat{R}_w^{(k+1)}(f) = (1-\alpha)\widehat{R}_w^{(k)}(f) + \alpha R_w(f), \tag{5.5}$$

where $\widehat{R}_w^{(k)}(f)$ is the estimate of the reward function at frequency f at interval k, α is the learning parameter and $R_w(f)$ is the current evaluation of the reward function.

It is important to note that the AR-filter updates the estimates, but in fact does not specify what actions should be taken. Instead of greedily optimizing this reward function, arbitrary experimentation is allowed. This is an important property in a time varying environment and allows decoupling the learning phase from the decision policy.

5.6.3 Exploration Algorithms

The algorithms presented in Sect. 5.5 suffer from hardware complexity and extra energy cost due to out-of-band scanning. In this section, out-of-band scanning is not performed any longer. Hence, the candidate set, C_i, covers all IEEE 802.15.4 channels. Differentiation between the channels is based on the observed history, rather than instantaneous observations.

The algorithms presented in Sect. 5.5 also suffer from over-exploration when steady-state has reached, as no cooling scheme is defined. As mentioned above, the goal of simulated annealing is to balance exploitation and exploration by defining a cooling scheme to define the value and evolution of T in Eq. 5.4. When T is high, the channel is selected randomly. However, as we lower T according to some cooling scheme, exploration is reduced and eventually we converge to a greedy selection. This means that the DT procedure used in this chapter is the simple RFS.

However, we show in Sect. 5.6.3.2, why a blind learner will perform very suboptimally in the present context and how the framework presented in Chap. 3 can be used to overcome this problem. Hence, in the present context DT knowledge is not

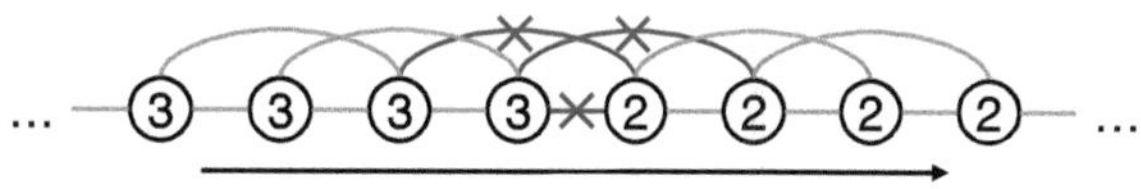

Fig. 5.4 Local optima cause very large delays. Cooling down in a local optimum causes a disjunction between two sets of terminals that have converged on different channels. Any packet generated on the left-hand side will never reach the sink

used to increase convergence speed or performance while converging. It is used to increase the performance of the steady-state result.

As in Sect. 5.5, we first consider a reward-based exploration instantiation of simulated annealing. In Sect. 5.6.3.2, we then explain the obstacles faced when designing a cooling scheme. Section 5.6.3.3 then explains how we have overcome these obstacles using a scenario-based framework, as the one presented in Chap. 3.

5.6.3.1 Reward-Based (RB) Exploration

In this section, the temperature, T, is kept constant and we do not define a cooling scheme. Hence, the rate of exploration is only altered through the reward, $\widehat{R}_w^{(k)}$. If the reward of the current state is high, less exploration is allowed. However, since no cooling scheme is defined, we will never eliminate exploring and terminals can still jump out of the optimal solution once in a while. We have observed that this effect actually can strongly reduce the efficiency (delay) in the steady-state. It is therefore necessary to find a cooling scheme.

5.6.3.2 Finding a Cooling Scheme

Finding a proper cooling scheme is a difficult task. If we just anneal and gradually drop the temperature to 0 K, we risk getting stuck in local optimum or are not able to adapt to the time-varying IEEE 802.11 interference. In many problems a trade-off is made between the speed of convergence and attaining the global optimum. However, the local optima of our derived problem are extremely suboptimal in the original problem statement (see Sect. 5.4)!

In Fig. 5.4 we illustrate this: a local optimum for the approximated goal destroys every opportunity to get the packet to the sink. Though exploring terminals might be able to bridge the packet over the boundary, a system that has been cooled down cannot (as no exploration is anymore allowed). Therefore reaching an optimal solution can not be traded off with convergence speed. This makes the selection of a proper cooling scheme (if there is any at all) very difficult and we have to make for each scenario a very wise choice of all parameters involved. Even then, it can still happen that we get stuck in a local optimum. In the next section, we will discuss how we can use the framework presented in Chap. 3 to design a more robust algorithm.

5.6.3.3 Adaptive Simulated Annealing

The high-level idea of our approach is similar to Adaptive Simulated Annealing (ASA). As discussed, removing exploration increases the chance of being stuck in a local optimum. Furthermore, it removes the ability of the framework to respond to dynamism.

However, a first scenario that we consider is the appearance of IEEE 802.11 interference on the current channel. When such interference is detected, the exploration is reinitialized, i.e., we reinitialize the temperature. As communication is no longer possible under IEEE 802.11 interference, this can be easily detected. Furthermore, when the environment suddenly changes, e.g., lots of neighbors disappear, one should also re-initialize the temperature. This allows the framework to perform optimally, while still being able to respond to the environment dynamism.

As mentioned in Sect. 5.6.3.2, we also need to protect the network against local optima. Our approach is rather similar to what we described above. Again, we try to detect if we are in a local optimum and increase exploration rate around the boundary. Hence, we see convergence centers created at the boundaries that ripple through the entire network chain. It is only when the different converged regions will agree on the channel that exploration is completely given up.

The detection algorithm now is not so easy anymore as in the interference case. As mentioned above, we have to rely on routing information. As it is known to the terminals whether other terminals are further or closer to the sink (in the routing table), we can use this information to detect boundaries. At a boundary a terminal will see a lot of terminals either closer to the sink or farther from the sink. If a boundary is detected, the boundary algorithm is initialized. It should be noted that the detection algorithm is only executed when the terminals have cooled down (after they have spent more than x iterations on a temperature lower than T_c). Otherwise, this detection algorithm would make a decision too quickly. This would interfere with the AR algorithm and can cause non-convergence.

We will now detail the boundary algorithm. The boundary terminals re-initialize the temperature, going for heavy exploration. Furthermore, boundary terminals are not allowed to re-enter the previous channel for some time. The boundary algorithm is hence a complete exploration phase. There are two possibilities:

- One side of the boundary has been quicker to converge. They will detect the boundary sooner and begin exploration. As a result, they will go looking for the terminals on the other side of the boundary. If they find the other terminals, new boundary terminals are formed in the quicker converged center. These terminals will then go to the same channel as the previous terminals and, hence, one side of the boundary adopts the channel of the other (1-sided ripple).
- Both sides of the boundary detect the boundary at almost the same time. The boundary terminals will converge on a third channel. At both sides new boundary terminals are formed and the new channel is rippled through both convergence centers (2-sided ripple).

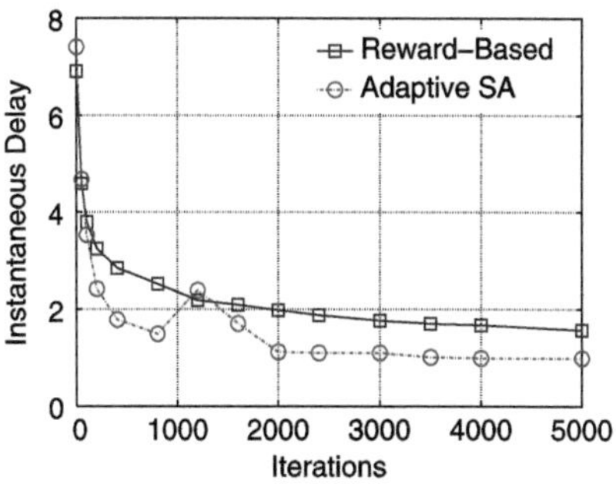

(a) Normalized delay for a string of 50 nodes ($R = 10$).

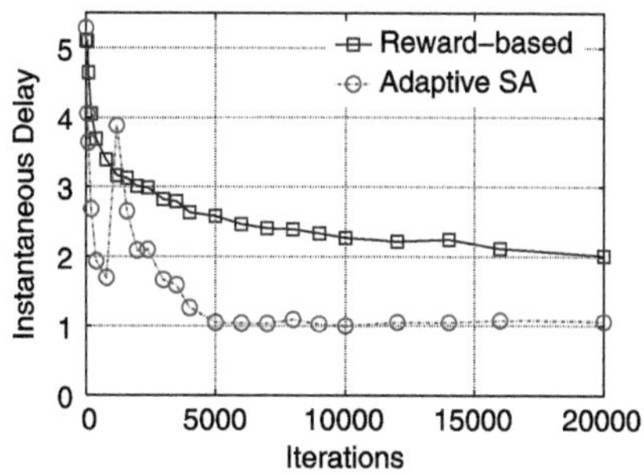

(b) Normalized delay for a string of 100 nodes ($R = 10$).

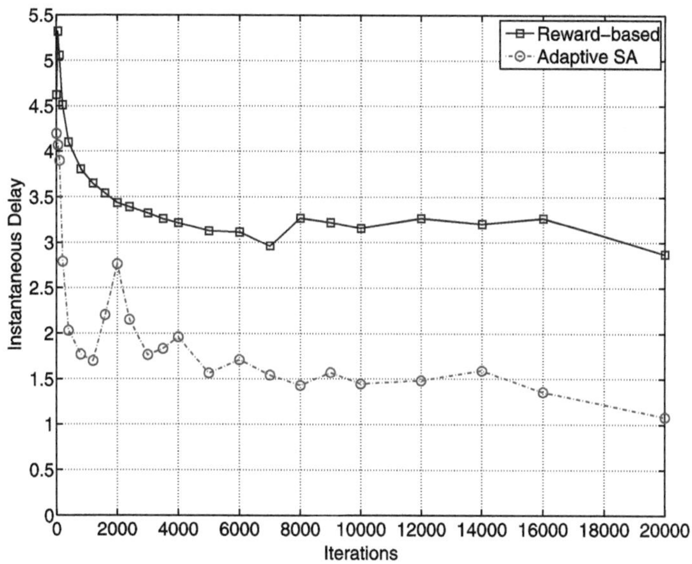

(c) Normalized delay for a string of 200 nodes ($R = 10$).

Fig. 5.5 These are instantaneous delays for different size networks. All 16 ZigBee channels are assumed to be available. Simulation procedure is the following: we allow the algorithms to reach a certain point. At that point, a packet is generated in the network. The instantaneous delay is then the delay of this packet. In these figures, the delays are averaged over 400 packets per run and 100 runs are executed

5.7 Simulation Results

In this section, we present some simulation results. In Fig. 5.5 the instantaneous delay for different situations is shown. Here, all 16 channels are assumed to be free of IEEE 802.11 interference. It can be clearly seen that the ASA algorithm outperforms the RB algorithm. Not only does it reach an optimal steady state (no exploration and all nodes on the same channel), it also converges significantly faster.

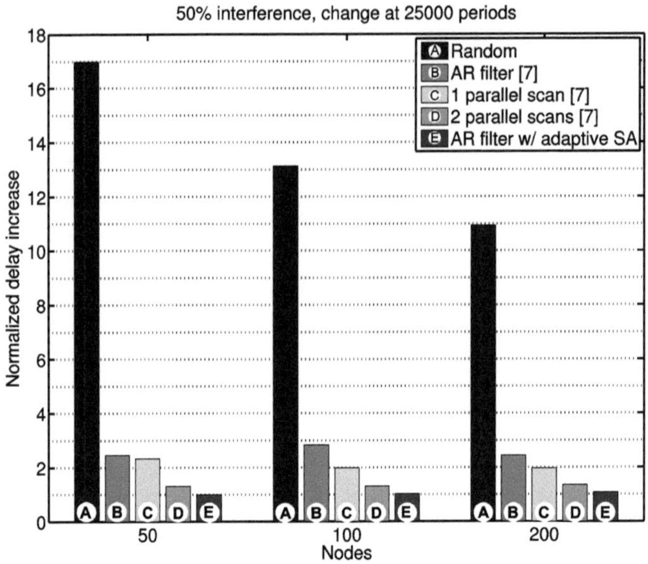

Fig. 5.6 Normalized delay increase compared to ideal channel allocation

However, we can see some spikes in Fig. 5.5. This is due to the local optima as discussed in Sect. 5.6.3.2. Initially, some neighboring nodes settle quickly on a channel but enough nodes remain exploring to bridge possible boundaries. Later on, these bridging nodes settle down and the algorithm gets stuck in a local optimum. As described in Sect. 5.6.3.3, the local optimum is detected and new convergence centers are rippled through the network. This increases the range of the converged areas, while eventually all nodes will settle down on one channel. Obviously, the bigger the network size, the slower the convergence.

For reference, we also present the results from a scenario presented in [65]. In Fig. 5.6 we can clearly see that the ASA algorithm reaches an almost optimal solution, i.e. a normalized delay of 1, even in the presence of a dynamic IEEE 802.11 interference. The results are even better than might be expected from Fig. 5.5. This is due to the dynamics of the IEEE 802.11 interference. As the interference is easily detected and degrades all communication, the IEEE 802.11 interference effectively reduces the available channels from 16 to 8. The presence of less available channels increases convergence speed. In the case that only one channel would be free for all nodes, any interference-aware algorithm will converge almost directly on that channel.

5.8 Conclusions

In this chapter, we have proposed fully distributed robust channel selection algorithms for IEEE 802.15.4 networks in the presence of dynamic IEEE 802.11 interference. The objective in this chapter was to minimize the end-to-end delay from packets generated randomly in a large network to a sink.

First, we restated the problem by inspecting the optimal solution under low-throughput assumption. We have shown that in the optimal solution all nodes use the same channel. Hence, we have redefined our goal function from minimizing delay to getting all the terminals on the same channel.

We have proposed both scan-based approaches that use instantaneous information and scanless-based approaches that learn the environment through past observations. These algorithms improve over those presented in [65] by letting the nodes settle down and freeze in a certain state. Due to the reformed problem statement, we have shown that this is only possible using an adaptive scenario-based framework, as presented in Chap. 3.

Simulation results have shown that all algorithms are able to react to the dynamic nature of the IEEE 802.11 interference. However, detailed delay analysis has proven that our scanless approach outperforms the scan-based proposals, while also having a lower hardware complexity and energy cost.

Due to the reformed problem statement, the developed algorithms are naturally suited to find a Common Control Channel (CCC) for an OSA network.

Chapter 6
Anticipative Energy and QoS Management: Systematically Improving the User Experience

6.1 Energy Efficiency for Smart Radios

6.1.1 Minimum Energy at Sufficient QoS

While the need for flexibility and intelligent control is well studied in the context of smart or cognitive radios for the sake of spectrum efficiency, it can used for optimizing the usage of other scarce resources such as energy. The design framework for smart radios of Chap. 3 is instantiated here to achieve smart, energy efficient software radios. More specifically, it is shown how it can be used to achieve smart, energy efficient IEEE 802.11a devices, provided their hardware is designed to be energy scalable. An energy scalable radio can be reconfigured or tuned at run-time in a way that energy consumption is impacted, so such a radio can benefit from a smart control strategy to save energy. Of course, this energy saving should not come at the expense of the user experience, so the objective of the smart control in this chapter is: how to adapt the radio so that the application performance is met at minimal possible energy consumption.

The focus of this chapter is on wireless networks where all users are in the same collision domain with an access point (AP) to arbitrate exclusive channel access (Fig. 6.2). An uplink real-time video streaming application is considered. The challenge in energy management for these systems is to determine how the system should sleep or scale and still meet per-packet QoS timing requirements. The instantiation takes advantage of the fact that the system will operate in dynamic environments where a single energy management solution is not sufficient. The main dynamics are encountered in the channel state and application load requirements. Furthermore, in a network of such systems, an efficient energy management algorithm should exploit the variations across users to minimize the overall network energy consumption.

Therefore the problem explored here could be stated as follows: *How does one decide what system configurations to assign to each user at run-time to minimize the overall energy consumption while providing a sufficient level of QoS? This must*

S. Pollin et al., *Software Defined Radios*, Signals and Communication Technology,
DOI 10.1007/978-94-007-1278-2_6,

be achieved for a network of users with bursty delay-sensitive data and over a slow fading channel.

We briefly recapitulate the following three observations that show that the proposed adaptation problem is not trivial and requires to streamline energy management approaches across layers.

- First, state-of-the-art wireless systems such as 802.11a devices are built to function at a fixed set of operating points assuming worst-case conditions. Irrespective of the link utilization, the highest feasible transmission rate is used and the power amplifier operates at maximum output power [21]. For non-scalable systems, the highest rate results in the smallest duty cycle and hence the lowest energy consumption. On the other hand, for scalable systems, this strategy results in excessive energy consumption for average channel conditions and link utilizations. Recent energy-efficient wireless system designs focus on VLSI implementations and adaptive physical layer algorithms where a lower transmission rate results in energy savings per bit [38, 71]. For these schemes to be practical, they should be aware of the hardware energy characteristics at various operating points.
- Second, to realize significant energy savings, systems need to shutdown the components when inactive. This is achieved only by tightly coupling the MAC communication to the power management strategy in order to communicate traffic requirements of each user for scheduling shutdown intervals.
- Finally, intrinsic trade-offs exist between schemes while satisfying the timeliness requirements across multiple users. As the channel is shared, lowering the rate of one user reduces the time left for the other delay-sensitive users. This forces them to increase their rate, at the cost of energy consumption or bit errors.

Our approach couples these tasks in a systematic manner to determine the optimal system-wide power management at run-time.

6.1.2 Smart Aspects and Energy Efficiency

We propose a two-phase run-time/design-time solution to efficiently solve the sleep-scaling trade-off across the physical, link and MAC layers for multiple users. It is an instantiation of the abstract design flow discussed in Chap. 3. At design-time the problem is resolved by searching for a set of close-to-optimal points in the solution space. This anticipates a set of possibly good system configurations. Starting from this configuration space, we can schedule the nodes at run-time to achieve near-optimal energy consumption with low overhead. The flow is illustrated in Fig. 6.1. It is clear that the set of possible points is fully characterized at design time. At run time, as function of the monitored scenario, the most energy efficient configuration can then be selected. In the next section, the design approach is detailed further. The research in this chapter has been performed in close collaboration with Rahul Mangharam at Carnegie Mellon University. This cooperation also leads to several shared papers [72–75].

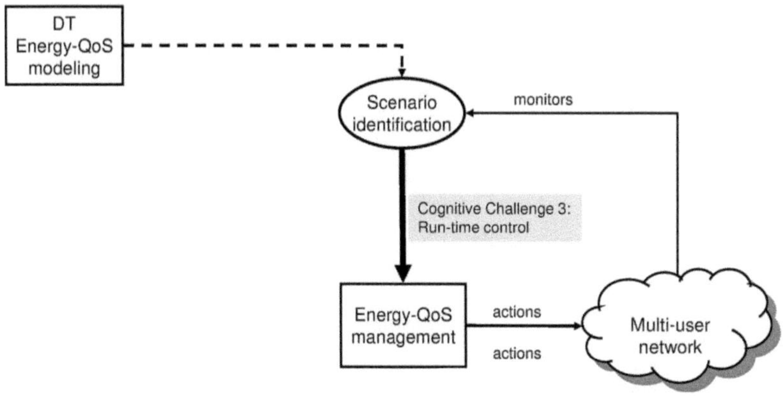

Fig. 6.1 Focus of this chapter is on the run time control challenge for cognitive radio. Actions are taken based on a monitoring of the environment (channel and application state) and based on this information the optimal configuration point is selected based on a DT model

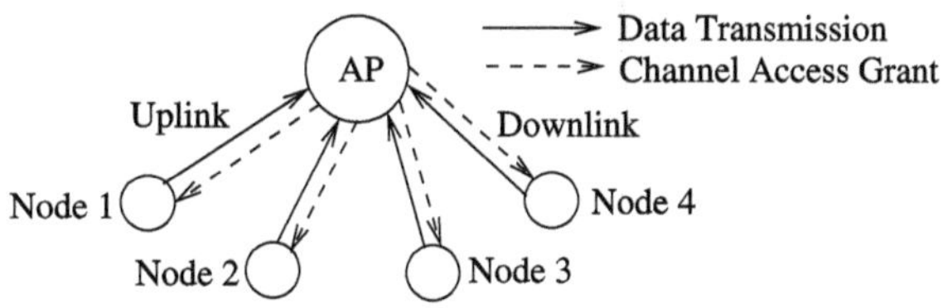

Fig. 6.2 Centrally controlled point-to-multipoint LAN topology with uplink and downlink communication

6.2 Anticipation Through Design Time Modeling

In this section, we follow the design steps considered in the design flow proposed in Chap. 3 to achieve the required performance at minimal QoS for the cognitive radios. Consider a wireless network as in Fig. 6.2 where multiple nodes are centrally controlled by an AP. Each node (such as a handheld video camera) desires to transmit or receive frames in real-time and it is the AP's responsibility to assign channel-access grants. The resource allocation scheme within the AP specifies each user's system configuration settings for the next transmission based on the feedback of the system state from the current transmission. It must ensure that the nodes meet their performance constraints by delivering their data in a timely manner while consuming minimal energy. The problem is first stated formally and a specific example is provided further in this chapter. The network consists of n flows $\{F_1, F_2, \ldots, F_n\}$ with periodic delay-sensitive frames or jobs. For notational simplicity, we assume a one-to-one mapping of flows to nodes, but the method is also applicable to more flows per node.

The design time steps, the focus of this chapter, are discussed in the next subsections. First, it is required to define the control dimensions or the available flexibility. Next, the system scenarios need to be determined or what scenarios the cognitive radio is expected to encounter. Then, the relevant cost, resource and quality dimensions need to be defined. They define what the cognitive radio should adapt for.

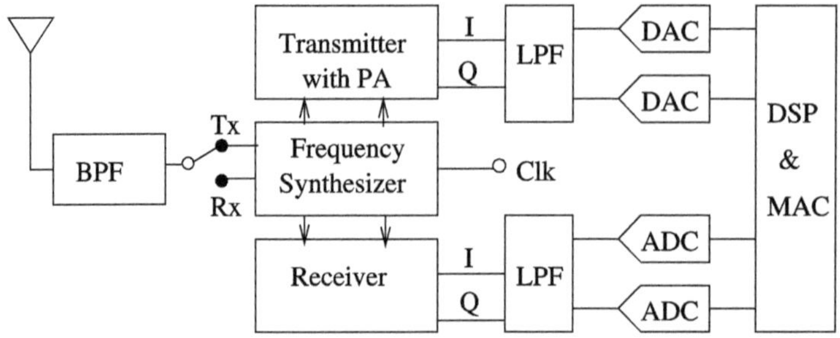

Fig. 6.3 802.11a OFDM direct conversion transceiver

These cost, quality and resource dimensions can be layered, to facilitate run-time information flow and ease of modeling. Finally, the mapping of the control dimensions impact on cost, quality and resource needs to be determined, for each of the system state scenarios. This means that for each possible scenario that the cognitive radio would encounter, it should be determined what the impact of a certain control know configuration would be on the cost, quality or resource dimensions. After these four design-time phases, the run-time phase can be carried out, as will be explained in Sect. 6.3.2.

6.2.1 Flexibility for Energy and QoS

Control Dimensions or Knobs ($K_{i,j}$): For the considered 802.11a transceiver, we identify several control dimensions that trade-off performance for energy savings. Our system modeling is based on an 802.11a [76] direct conversion transceiver implementation with turbo coding [77] (Fig. 6.3). Four control dimensions have a significant impact on energy and performance for these OFDM transceivers: the modulation order (N_{Mod}), the code rate (B_c), the power amplifier transmit power (P_{Tx}) and its linearity specified by the back-off (b).

For the energy modeling, we focus on the power amplifier control knob as PA's generally are the most power-hungry component in the transmitter consuming upwards of 600 mW [78]. The major drawback for 802.11a OFDM modulation is the large peak-to-average power ratio (PAPR) of the transmitted signal.[1] A high PAPR renders the implementation costly and inefficient since power-efficient PA designs require a reduced signal dynamic range [80]. However, reducing the PA's linear power amplification range clips the transmitted signal and increases the signal distortion.

To combat this, new PA designs allow to adapt the gain compression characteristic independently of the output power. Adapting the PA gain compression characteristic allows to translate a transmit power or linearity reduction into an effective

[1]We note that digital techniques such as clipping also enable to significantly decrease the PAPR constraint [79].

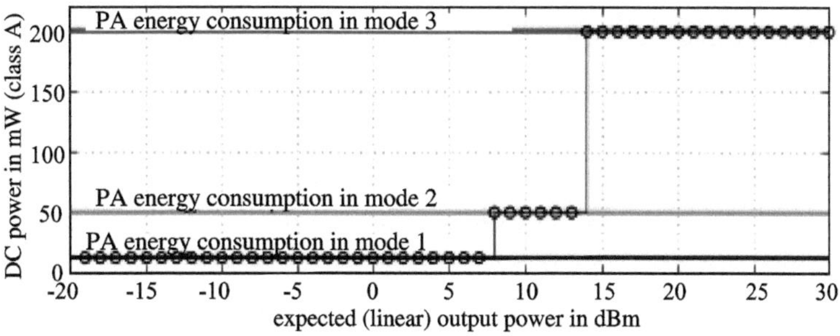

(a) Effective power consumption.

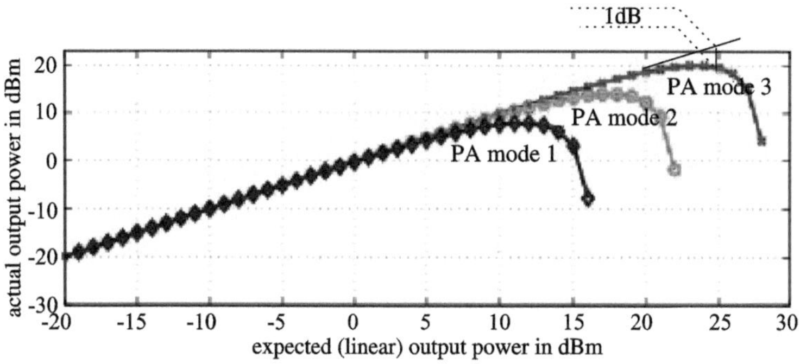

(b) Linearity and 1dB compression point.

Fig. 6.4 Adapting the PA gain compression characteristic allows to translate a transmit power or linearity reduction into an effective energy consumption gain

energy consumption gain. A back-off, b, can be used to steer the linearity of the amplification system versus the energy efficiency of the PA [21]. The back-off is defined as the ratio of the effective PA output power to the saturation output power corresponding to the 1 dB gain compression point (Fig. 6.4(b)). The saturation power and hence signal distortion for class A amplifiers (used with OFDM) are controlled by modifying the bias current of the amplifier, which directly influences its energy consumption (Fig. 6.4(a)). Since the linearity of the PA determines the distortion added to the signal, we can write the performance loss due to the PA as function of b: $D_i(b)$. Consequently, we save energy from the increased PA efficiency, provided we ensure that the received signal to noise and distortion ratio (SINAD) is above the required sensitivity and do not need to retransmit the packet. For the system to be practical only discrete settings of the control dimensions are considered (listed in Table 6.1).

We further consider the eight PHY rates supported by 802.11a based on four modulation and three code rates (Table 2.2). Given the above introduced N_{Mod} and B_c, the bitrate (B_{bit}) achieved for each modulation-coding pair with N_c OFDM carriers and symbol rate B is given by:

$$B_{bit} = N_c \times N_{Mod} \times B_c \times B. \tag{6.1}$$

Based on the bitrate, communication performance is determined by the Bit Error Rate (BER) at the receiver. When transmitter non-linearity is considered, the BER is expressed as a function of the above introduced SINAD. The SINAD is written as a function of the power amplifier back-off, given output power P_{Tx} and time and frequency varying channel attenuation $A(t, f)$ as:

$$\text{SINAD} = \frac{P_{Tx} \times A(t, f)}{A(t, f) \times D_i(b) + kT \times W \times N_f}, \tag{6.2}$$

$$P_{PA} = \frac{P_{Tx}}{\eta_{PA}(b)}, \tag{6.3}$$

where the constants k, T, W and N_f are the Boltzmann constant, working temperature, channel bandwidth and noise figure of the receiver respectively. The relation between the power amplifier back-off b and the distortion has been characterized empirically for the Microsemi LX5506 [81] 802.11a PA in Fig. 6.4. The effective PA power consumption (P_{PA}) can be expressed as the ratio of the transmit power (P_{Tx}) to the PA efficiency (η_{PA}) that is related to b by an empirical law fitted on measurements (Eq. 6.3).

Next, at MAC layer, we can add the number of retransmissions as control knob. These knobs are practical and sufficient to illustrate the potential of the proposed methods.

6.2.2 The Varying Context

System State ($S_{i,m}$): In a wireless environment, with e.g. VBR video traffic, the system state scenarios used in this design case are the current channel state and application frame size. We study the impact of both constant bitrate (CBR) and variable bitrate (VBR) traffic. For VBR traffic we employ MPEG-4 encoded video traces [82, 83] with peak-to-mean frame sizes ranging from 3 to 20. All fragmentation is done at the link layer and we use UDP over IP. As the maximum frame size is assumed to be within the practical limit of 50 fragments, we construct Cost-Resource-Quality curves for 1, 2, 3, 4, 5, 10, 20, 30, 40, 50 fragments/frame and interpolate for intermediate values. The application layer frame size is translated to lower layer Queue Size and fragment size to facilitate state monitoring and calibration. As a result, no additional measurements are needed to model traffic requirements, which can be fully captured in the mapping.

The indoor wireless channel suffers from frequency-selective fading that is time-varying due to movements of the users or obstacles in its environment. This varying fading results in varying PER as function of the current transceiver settings. We use a frequency selective and time varying channel model to compute this PER for all transceiver settings. An indoor channel model based on HIPERLAN/2 [84] was used for a terminal moving uniformly at speeds between 0 to 5.2 km/h (walking speed). Experiments for indoor environments [83] have found the Doppler spread to be approximately 6 Hz at 5.25 GHz center frequency and 3 Hz at the 2.4 GHz center

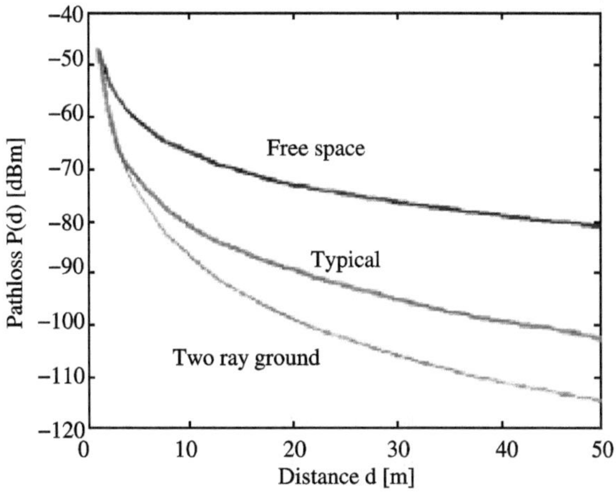

Fig. 6.5 Typical indoor pathloss model

frequency. This corresponds to a coherence time of 166 ms for 802.11a networks. A set of 1000 channel frequency response realizations (sampled every 2 ms over one minute) were generated and normalized in power. Data was encoded using a turbo coder model [77] and the bit stream was modulated using 802.11a OFDM specifications. For a given back-off b and transmit power P_{Tx}, the SINAD at the receiver antenna was computed for the channel realization $A(t, f)$ by Eq. 6.2.

We assume an average path-loss A of 80 dB at a distance of 10 m, which consists both of the transmit antenna gain G_t, receive antenna gain G_r and the propagation loss. This path-loss is from an empirical model developed at IMEC (Fig. 6.5). It is representative for indoor channels and results in pathlosses in between the *Free space* model and *Two ray ground* model, which is commonly used for outdoor applications.

From the channel realization database, a one-to-one mapping of SINAD to receive block error rate was determined for each modulation and code rate. The channel is then classified into 8 classes, determined by a 2 dB difference in the receive SINAD that is measured for a turbo code BlER of 10^{-3} (Fig. 6.6(a)). We use a similar 2 dB discrete step for the PA knobs (Table 6.1). In order to derive a time-varying link-layer error model, we associate each channel class to a Markov state, each with a probability of occurrence based on the channel realizations database (Fig. 6.6(b)). Given this eight-state error model, we are able to model the PER, expressing the performance per MAC layer packet L_{frag}, for different configurations. The PER is obtained in Eq. 6.4 by assuming the block errors are uncorrelated for a packet size of L_{frag} bits and a block size of 288 bits:

$$P_e = \left[1 - (1 - \text{BlER})^{\frac{L_{frag}}{288}}\right]. \tag{6.4}$$

This calibration is not a fundamentally new characterization of the hardware. This BlER-SINAD information just has to be combined with a set of channel states that are extracted on-line. The combination of this run-time channel with hardware properties into metrics for higher layer metrics of the database is a key enabler.

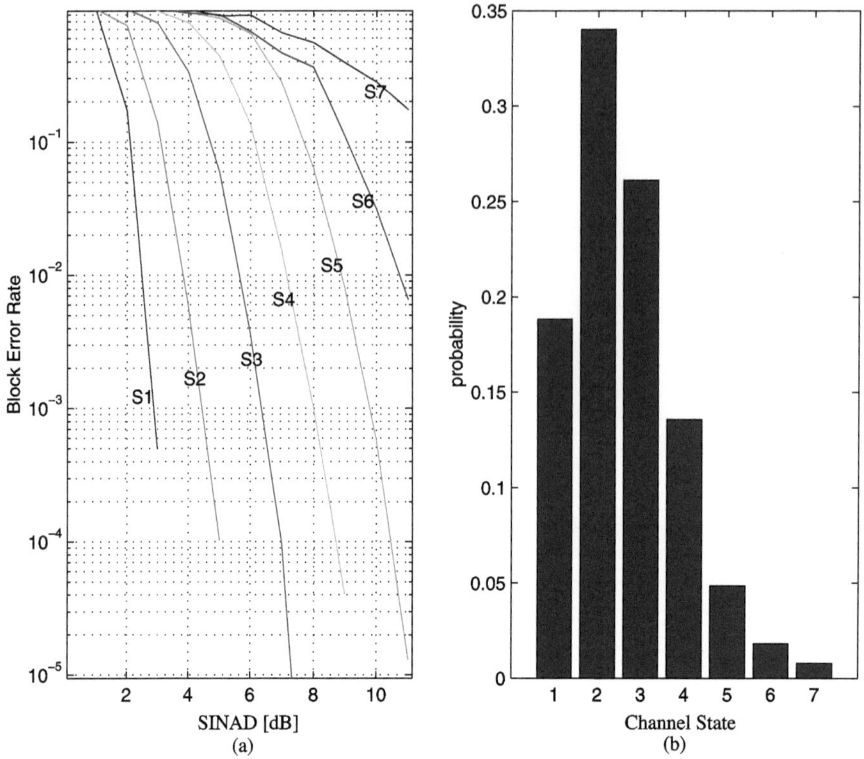

Fig. 6.6 Markov channel model used for indoor 802.11a wireless communication (**a**) BlER versus SINAD and (**b**) histogram for steady state Markov state probabilities

6.2.3 Objectives for Efficient Energy and QoS Management

We determine the set of quality, energy cost and resource dimensions that are relevant to the considered scenario where users transmit video data towards a central AP (Fig. 6.2).

1. **Cost Function** (C_i): The optimization objective considered in the design case in this chapter is to minimize the total energy consumption of all users in terms of Joules/Job. For example, in a video context, a job is the timely delivery of the current frame of the video application.
2. **QoS Constraint** (Q_i): The optimization has to be carried out taking into account a minimum performance or QoS in order to satisfy the user. As delivery of real-time traffic is of interest, we describe the QoS in terms of a single QoS metric which is in this design case defined to be the job failure rate (JFR) [85]. JFR is defined as the ratio of the number of frames not successfully delivered before the deadline to the number of frames issued by the application over the lifetime of the flow. In this specific delay-sensitive video context, the QoS constraint is hence specified at run-time by user i as a target JFR_i^*. This QoS definition could

be extended to a broader range of quality targets. We note again that, next to this explicit QoS constraint that can be adapted or configured at run-time, application bitrate constraints are implicit constraints given through by the use of scenarios.

3. **Shared Resource** (R_i): In this dissertation, we consider the particular case where access to the channel is only divided in time. The fraction of resource consumed by node i is denoted by R_i.

To summarize, each flow F_i is associated with a set of possible system states $S_{i,m}$, which affects the mapping of the control dimensions $\mathbf{K}_i$ to the Cost ($\mathbf{K}_i, S_{i,m} \rightarrow C_i$), Resource ($\mathbf{K}_i, S_{i,m} \rightarrow R_i$) and Quality ($\mathbf{K}_i, S_{i,m} \rightarrow Q_i$) that will be specified in the next section. It is essential to note that for each user, depending on the current state, the relative energy gains possible by rate scaling or sleeping are different and should hence be exploited differently. Each user experiences different channel and application dynamics, resulting in different system states over time, which may or may not be correlated with other users. This is a very important characteristic which makes it possible to exploit multi-user diversity for energy efficiency.

As the system performance requirements are specified at the application layer and the energy consumption is at the lower (hardware) layers, it is essential to: (a) translate the application layer requirements to relevant metrics at each intermediate layer and (b) to define clean interfaces between layers for an Energy-Performance feedback mechanism (see Fig. 6.7). This is to allow for a local calibration of the hardware which makes implementation more feasible, still enabling translation of to application specific quality metrics. For delay-sensitive traffic, the QoS metric of interest is the target JFR^*.

At the link-layer, each application *frame* is fragmented into one or more fixed-sized *fragments*. An application frame size or rate requirements are hence translated into a local Queue Size. Next, as shown in Fig. 6.7, the JFR^* is translated at the link layer to a PER constraint, which corresponds to a maximum Block Error Rate (BlER) as a function of the physical layer low-level knobs. The BlER is a result of the receive SINAD for given PHY parameters. Each target JFR may be satisfied by one or more control dimension configurations, $\mathbf{K}_i$, each associated with the energy consumed (cost) and time required (resource) to complete the frame transmission in the current system state. The state is defined by a discrete channel state and traffic requirement (i.e. current frame size), which can easily be monitored as the Queue Size. Channel classification and monitoring is typically a more difficult problem. At run-time, based on a node's current system state and JFR^*, the corresponding Cost-Resource set of points are fetched from memory. From each curve, configuration settings are then chosen using the fast greedy algorithm such that the total transmission time for all nodes is less than the deadline. We first discuss the control dimension mapping to obtain those Pareto-optimal trade-off sets.

6.2.4 Anticipating the Performance

The key aspects of the method for design of an energy efficient cognitive radio are the mapping of the control dimensions to cost, resource and quality profiles

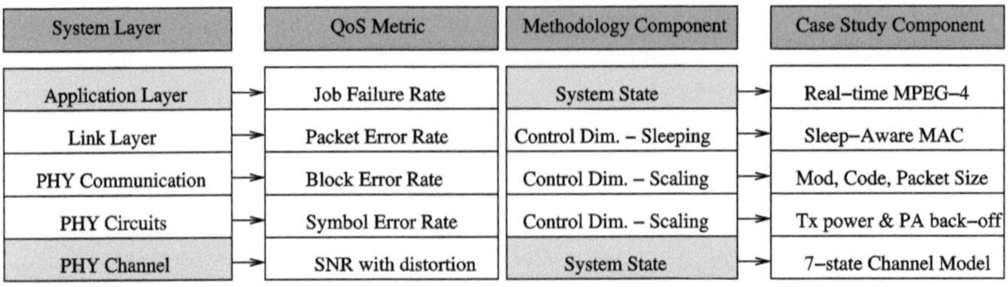

Fig. 6.7 A smart radio design approach spans multiple layers with corresponding performance metrics. The case study demonstrates the energy management methodology in the 802.11a WLAN setting

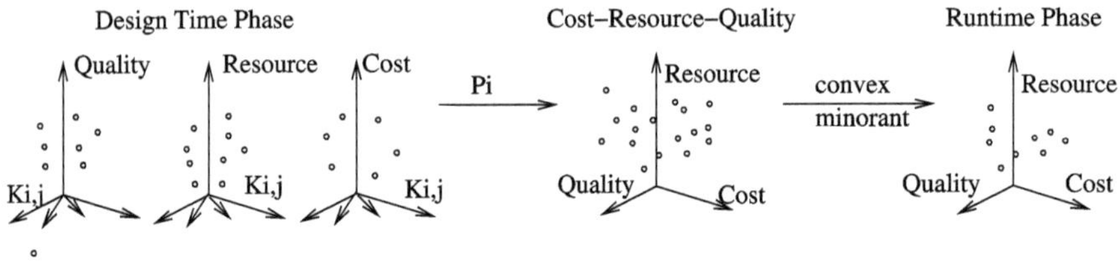

Fig. 6.8 At design-time, a Cost, Resource and Quality profile is determined for each set of control dimensions based on the system state. The optimal Cost-Resource-Quality trade-off is derived from this mapping to give operating points used at run-time

respectively, and the generality of this mapping. A resource (respectively cost, quality) profile describes a list of potential resource (respectively cost, quality) allocation schemes resulting from each configuration point. These profiles are then combined to give a Cost-Resource-Quality trade-off, which is essential for solving the resource allocation problem (Fig. 6.8). The Cost-Resource-Quality trade-off function represents the behavior of a specific system for one user in a given system state.

Cost profile properties

- The finite set of discrete control dimension configurations can be ordered by their increasing cost.
- The overall system cost, C_{net}, is defined as the weighted sum of costs of all flows, where each flow can be assigned a certain weight depending on its relative importance or to improve fairness [86]. Users may be assigned higher weights for example when their battery capacity is low or when they downscale their transmission rate by decreasing the video quality and get rewarded for reducing the network congestion. Users with a higher weight will typically be allowed to save more energy compared to other users: $C_{net} = \sum_{i=1}^{n} \omega_i C_i$.

Resource profile properties

- The finite set of discrete control dimension configurations can be ordered according to minimal resource requirement.
- The system resource requirement, R, is defined as the sum of the per flow requirements: $R = \sum_{i=1}^{N} R_i$.

Quality profile properties

- The finite set of discrete control dimension configurations can be ordered with quality.
- The system quality, Q, is met when each individual user's constraint is met: $JFR_i \leq JFR_i^*$, $1 \leq i \leq n$.

The finite set of discrete control dimensions can be ordered, describing a range of possible costs, resources and quality for the system in each system state. For each additional unit of resource allocated, we only need to consider the configuration that satisfies the quality constraint and achieves the minimal cost for that resource unit. For each system state (e.g., channel and application loads), a subset of points is determined by pruning the Cost-Resource-Quality set of points to yield only the minimum cost configurations, which will be denoted by $C_i(R_i, Q_i)$.

We define a calibration function p_i, that is computed for every state $S_{i,m}$

$$p_i(R_i, Q_i) = \min\{C_i \mid (\mathbf{K}_i, S_{i,m} \rightarrow R_i) \wedge (\mathbf{K}_i, S_{i,m} \rightarrow Q_i) \\ \wedge (\mathbf{K}_i, S_{i,m} \rightarrow C_i) \wedge (\mathbf{K}_i) \in \{\mathbf{K}_i\}\}$$

and defines a mapping between the Resource, Cost and the Quality of a node in a system state, $S_{i,m}$, as shown in Fig. 6.8. $\{\mathbf{K}_i\}$ is the set of configuration vectors for node i. Given the points after calibration in the Cost-Resource-Quality space, we are only interested in the ones that represent optimally the trade-off between energy, resource and quality for our system. Although the discrete settings and non-linear interactions in real systems do not lead to a convex trade-off, it can be well approximated as follows.

We calculate the *convex minorant* [87] of these pruned curves along the Cost, Resource and Quality dimensions, and consider the intersection of the results. We call this set the optimal Cost-Resource-Quality trade-off in the remainder. The optimal Cost-Resource-Quality trade-off is plotted in the dimensions TXOP-JFR-Energy/frame in Fig. 6.9(a) for the considered 802.11a WLAN design case. We will show later that the maximum segment size of the convex minorant determines the solution's deviation from the optimum.[2] Configurations on segments that are small compared to the largest segment size can be pruned away without affecting the bounds of the solution. As a result, we can typically expect less than 30 configurations per state. At run-time, the resource allocation scheme in the AP adapts to

[2] To achieve an optimum, it is necessary to retain the set of points that are Pareto-optimal or dominant in the Cost-Resource-Quality dimensions. A complex optimization problem with backtracking has to be solved at run-time to achieve the optimum based on the Pareto-points.

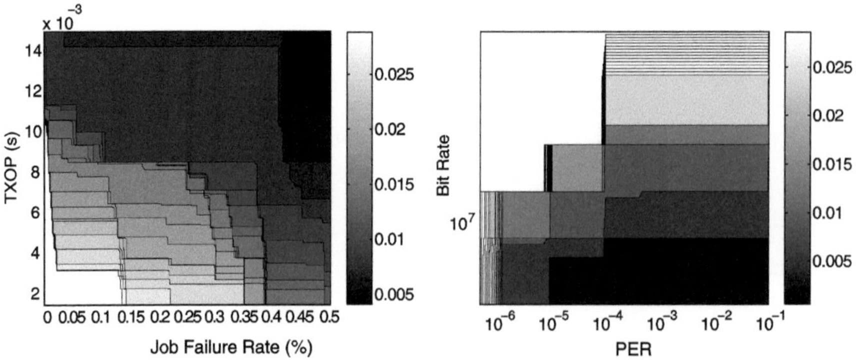

(a) MAC layer. (b) Physical layer.

Fig. 6.9 Control dimension mapping

the system state by fetching the correct configurations from memory. This operation is cheap compared to the cost of calibration that only has to be carried out once. In the next subsection we detail how this information can be combined efficiently into a global set representing the network trade-off. Profiling each user separately and combining the information at run-time is optimal for independent users or when the correlation is unknown. When correlation is present and known, the number of system states to calibrate and the run-time combination could be further reduced.

6.3 Managing the User Experience

We first formally state the energy efficient cognitive radio resource allocation problem to be solved at run-time. Then we propose a very efficient algorithm to achieve the network-wide optimal configuration.

6.3.1 Smart Resource Allocation Problem Statement

We recall that our goal is to assign transmission grants via the AP, resulting in an optimal setting of the control dimensions to each flow and node such that the per-flow QoS constraint for multiple users are optimally met with minimal energy consumption. For a given set of resources, control dimensions and QoS constraints, the scheduling objective can be formally stated as:

$$\min_{C_{net}} \sum_{i=1}^{n} \omega_i C_i, \quad 1 \leq m \leq s \tag{6.5}$$

s.t.

$$JFR_i \leq JFR_i^*, \quad 1 \leq i \leq n \qquad \text{(QoS Constraints)}, \tag{6.6}$$

$$\sum_{i=1}^{n} R_i \leq R^{max} \qquad \text{(Resource Constraints)}, \tag{6.7}$$

$$\mathbf{K}_i, S_{i,m} \rightarrow R_i, \quad 1 \leq m \leq s \qquad \text{(Resource Profiles)}, \tag{6.8}$$

$$\mathbf{K}_i, S_{i,m} \rightarrow C_i \qquad \text{(Cost Profiles)}, \tag{6.9}$$

$$\mathbf{K}_i, S_{i,m} \rightarrow Q_i \qquad \text{(Quality Profiles)}. \tag{6.10}$$

The solution of the optimization problem yields a set of feasible operating points, $\bar{\mathbf{K}} = \{\mathbf{K}_i, \ 1 \leq i \leq n\}$, for the network which fulfill the QoS target, maintain the shared resource constraint and minimize the system cost. While the profile mapping and pruning is done during a one-time calibration step, we now propose how the optimal configuration $\bar{\mathbf{K}}$ is determined efficiently at run-time using those profiles.

6.3.2 Greedy Resource Allocation

As the current system state of all the nodes and flows is only known at run-time, a light-weight scheme is necessary to assign the best system configurations for each user, while meeting the QoS requirements. We therefore employ a greedy algorithm to determine the per-flow resource usage R_i for each flow to minimize the total cost C_{net} while meeting the system constraints. The algorithm first constructs the optimal local Cost-Resource trade-off curve $C_i(R_i)$ by taking the optimal points in both dimensions that meet the run-time average quality constraint JFR^*. Next, the scheduler traverses all flows' two-dimensional Cost-Resource curves and at every step consumes resources corresponding to the maximum negative slope across all flows (taking into account user preference or weight if appropriate). This ensures that for every additional unit of resources consumed, the additional cost saving is the maximum across all flows taking into account the agreements made at admission time. We assume that the current channel states and application demands are known. The exchange of state information and operating points between nodes and the AP is obtained by coupling the MAC protocol with the resource manager as explained in the next section.

We determine the optimal additional allocation to each flow, $R_i > 0$, $1 \leq i \leq n$, subject to $\sum_{i=1}^{n} R_i \leq R$. Our greedy algorithm is based on Kuhn–Tucker [87]:

1. Allocate to each flow the smallest resource for the given state, R_{min}. By assumption, all flows are schedulable under worst-case conditions, i.e., $\sum_{i=1}^{n} R_{min} \leq R$.
2. Let the current normalized allocation of the resource to flow F_i be R_i, $1 \leq i \leq n$. Let the unallocated quantity of the available resource be R_{avl}.
3. Identify the flow with the maximum negative slope, $|C_i'(R_i)|$, representing the maximum decrease in cost per resource unit (i.e. moving right and downward the $C_i(R_i)$ convex minorant in Fig. 6.10). If there is more than one, pick one randomly. If the value of the minimum slope is 0, then stop. No further allocation will decrease the system cost further.

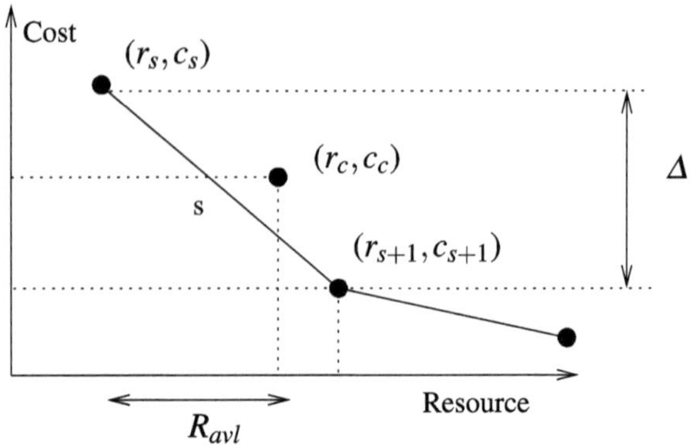

Fig. 6.10 Bounded deviation from the optimal in discrete Cost-Resource curves

4. Increase R_i by the amount till the slope changes for the ith flow. Decrement R_{avl} by the additional allocated resource and increment the cost C by the consequent additional cost. Return to step 3 until all resources have been optimally allocated or when R_{avl} is 0.

In our implementation, we sort the configuration points at design-time in the decreasing order of the negative slope between two adjacent points. The complexity of the run-time algorithm is $O(Ln\log(n))$ for n nodes (~ 20) and L configuration points per curve. Further in this chapter, we demonstrate that for a practical system in each possible system state (i.e., channel and frame size), the number of configuration points to be considered at run-time is relatively small (~ 30). Taking into account that the relation $C_i(R_i)$ is convex, we now prove that the greedy algorithm leads to the optimal solution for continuous resource allocation. Following that, we extend the proof for systems with discrete working points to show that the solution is within a bound from the optimal.

Theorem 4.1 *For a continuous resource allocation to be optimal, a necessary condition is* $\forall i,\ 1 \leq i \leq n,\ R_i = 0$, *or for any flows* $\{i, j\}$ *with* $R_i > 0$ *and* $R_j > 0$, *the cost slopes* $C_i'(R_i) = C_j'(R_j)$.

Proof For a continuous differentiable function, the Kuhn–Tucker [87] theorem proves such a greedy scheme is optimal. Suppose for some $i \neq j$, let the optimal resources be $R_i > 0$, $R_j > 0$, and $|C_i'(R_i)| > |C_j'(R_j)|$. As the savings in cost per resource for F_i is larger, we can subtract an infinitesimal amount of resource from F_j and add it to F_i. Total system cost is reduced, contradicting the assumed optimality. □

In a real system, however, the settings for different control dimensions such as modulation or transmit power are discrete. This results in a deviation, Δ, from the optimal resource assignment. We now show this worst-case deviation Δ from the optimal strategy is bounded and small.

Theorem 4.2 *$\exists 0 \leq \Delta < \infty$, such that $C_{OPT} \leq C_{MEERA} \leq C_{OPT} + \Delta$, where C_{OPT} is the optimal cost (energy consumed by all users) and C_{MEERA} is the cost in the discrete case.*

Proof For all flows, $\{F_1, F_2, \ldots, F_n\}$, the aggregate system resources consumed are stored in the decreasing order of their negative slope across all per-flow Cost-Resource $C_i(R_i)$ curves. Based on this ordering, the aggregate system $C(R)$ trade-off is constructed, consisting of segments resulting from individual flows. The greedy algorithm traverses the aggregate system $C(R)$ curve, consisting of successive additional resource consumptions (at maximum cost decrease), until the first segment, s, is found that requires more resources than the residual resource R_{avl} (Fig. 6.10).

Let the two end points of the final segment s be (r_s, c_s) and (r_{s+1}, c_{s+1}) in $C(R)$. Let (r_c, c_c) be the optimal resource allocation in the optimal combined Cost-Resource curve.

$$\begin{aligned} C_{OPT} &\geq C_{MEERA} - (r_c - r_s)(c_{s+1} - c_s)/(r_{s+1} - r_s) \\ &> C_{MEERA} - (r_{s+1} - r_s)(c_{s+1} - c_s)/(r_{s+1} - r_s) \\ &= C_{MEERA} - (c_{s+1} - c_s). \end{aligned}$$

We observe that $c_s - c_{s+1} \leq \Delta$, therefore $C_{MEERA} - C_{OPT} < \Delta$. Moreover, we note that with more dimensions $(K_{i,j})$ considered, a better approximation can be obtained. □

6.4 IEEE 802.11a Design Case

To demonstrate the usability of the proposed control scheme we apply it to control a flexible OFDM 802.11a modem. The target application is the delivery of delay-sensitive traffic over a slow fading channel with multiple users. We associate the system *Cost* to *energy*, the *Resource* to the *time* over the shared medium and the *Quality* is the *JFR*.

We briefly consider the trade-offs present across the physical circuits, digital communication settings and link layer in our system. Increasing the modulation constellation size decreases the transmission time but results in a higher PER for the same channel conditions and PA settings. The energy savings due to decreased transmission time must offset the increased expected cost of re-transmissions. Also, increasing the transmit power increases the signal distortion due to the PA nonlinearity [81]. On the other hand, decreasing the transmission power also decreases the efficiency of the PA. Considering the trade-off between sleeping and scaling, a longer transmission at a lower and more robust modulation rate needs to compensate for the opportunity cost of not sleeping earlier. Finally, as all users share a common channel, lowering the rate of one user reduces the time left for other delay-sensitive users. This compels other users to increase their rate and consume more energy or experience errors.

Table 6.1 PHY parameters considered

Performance Model	Energy Model	PHY Control Dimensions
$W = 20$ MHz	$P_{FE}^{T} = P_{FE}^{R} = 200$ mW	Back-off (dB) {6 to 16}
$B = 250$ kBaud	$E_{DSP}^{T} = E_{DSP}^{R} = 250$ nJ/symbol	P_{Tx} (dBm) {0 to 20}
$N_c = 48$	$E_{DSP}^{R} = 8.7$ nJ/bit	Modulation N_{Mod} $\{BPSK, QPSK, 16-QAM, 64-QAM\}$
Block = 288	$P_{idle} = 200$ mW	Code Rate B_c $\{1/2, 2/3, 3/4\}$
$N_f = 10$ dB	$L_{frag} = 1024$ Bytes	

6.4.1 Energy-Performance Anticipation

Four control dimensions for the physical layer have been introduced to steer energy and performance of these OFDM transceivers. As function of these control dimensions, the energy $E_{Tx}(\mathbf{K}_i)$ needed to send and the energy $E_{Rx}(\mathbf{K}_i)$ needed to receive a unit of L_{frag} bytes need to be characterized.

We realistically assume that the energy consumption of the digital baseband is a linear function of the number of symbols to decode (E_{DSP}^{R} and E_{DSP}^{T} in energy/symbol). For the turbo coder, the energy cost depends on the number of bits to decode (E_{DSP}^{R} in energy/bit) at the receiver [77]. The block size used for the turbo coding is 288 bits. Based on current implementations [78], the frequency synthesizer, Analog-to-Digital Converter (ADC), Digital-to-Analog Converter (DAC), Low-Noise Amplifier (LNA) and filters are assumed to have a fixed front-end power consumption P_{FE} as given in Table 6.1. The time needed to wake-up the system (stabilization time for the Phase-Locked Loop (PLL) in the frequency synthesizer) is assumed to be 100 μs, which is optimistic but can be achieved when designing frequency generators for this purpose. Application layer frames are fragmented at the link layer into L_{frag}-sized fragments. We obtain the following expressions for the energy needed to send or receive a fragment of length L_{frag}, as a function of the current knob settings:

$$E_{Tx} = \left(\frac{P_{PA} + P_{FE}^{T}}{B_{bit}} + \frac{E_{DSP}^{T}}{N_c \times N_{Mod} \times B_c} \right) \times 8L_{frag}, \tag{6.11}$$

$$E_{Rx} = \left(\frac{P_{FE}^{R}}{B_{bit}} + \frac{E_{DSP}^{R}}{N_c \times N_{Mod} \times B_c} + E_{DSP}^{R} \right) \times 8L_{frag}. \tag{6.12}$$

Finally, to complete the model, we introduce a term P_{idle} that denotes the power consumption when the transceiver is idle (Table 6.1).

Obtaining the actual values for energy consumption for each of the configurations only depends on the fragment size, and the analog and digital baseband power consumption values as function of the configuration settings. In practice, this information is obtained very fast by transmitting a L_{frag} packet once (requiring 0.1 to 1.3 ms depending on the configuration) using e.g. 1000 configurations (hence $1.3 \times 10^{-3} \times 1000 = 1.3$ s for a complete system profile). It is acceptable to spend

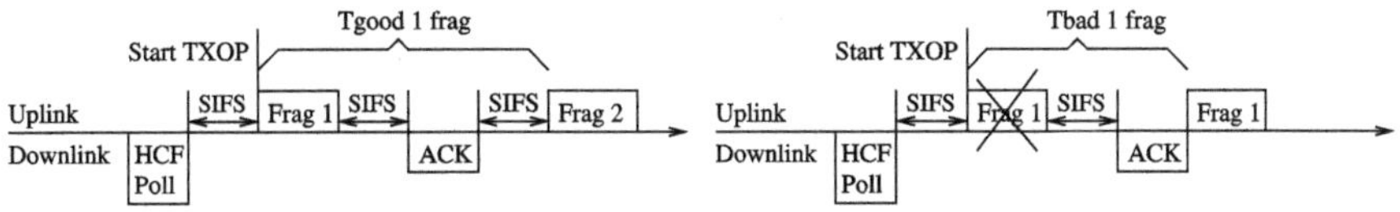

Fig. 6.11 Timing of successful and failed uplink frame transmission with 802.11e HCF

this time since this calibration only needs to be carried out once over the system lifetime and gives us sufficient information to fully characterizes the device's energy and performance as function of the control dimensions considered.

We now have expressions for the gross transmission rate (Eq. 6.1) and for the energy consumption to send L_{frag} bits. However, to obtain the complete energy and performance models, we have to also take into account all overheads and compute the total energy and resource cost for each scenarios.

For each combination of feasible control dimensions of node i, $\mathbf{K}_i$ (which we will simplify to $\mathbf{K}$), we compute the total expected energy consumption, total transmission time and resulting *JFR* while using the properties of a sleep-enabled 802.11e MAC. We assume that during each communication instance, all transmissions use the same configuration to eliminate reconfiguration costs. All transmissions employ the contention-free mode with transmit opportunity (TXOP) grants of 802.11e HCCA [88]. Let E_H, E_{ACK} and T_H, T_{ACK} be the pre-determined energy and time needed to transmit a header and acknowledgment. The energy and time needed for a successful E_{good} and failed[3] E_{bad} transmission is then determined following Fig. 6.11 and using parameters listed in Table 6.1:

$$E_{good}(\mathbf{K}) = E_{Tx} + E_H + (2T_{sifs}P_{idle}) + E_{ACK}, \tag{6.13}$$

$$E_{bad}(\mathbf{K}) = E_{Tx} + E_H + (T_{sifs} + T_{ACK})P_{idle}, \tag{6.14}$$

$$T_{good}(\mathbf{K}) = \frac{L_{frag} \times 8}{B_{bit}} + T_H + (2T_{sifs}) + T_{ACK}, \tag{6.15}$$

$$T_{bad}(\mathbf{K}) = T_{good}(\mathbf{K}) - T_{sifs}. \tag{6.16}$$

We now include the MAC layer retransmissions. Each fragment is transmitted with configuration $\mathbf{K}$, for which we can determine the Packet Error Rate P_e, based on Eq. 6.4. The probability that the frame is delivered with exactly $(m + p)$ attempts (including p retransmissions), is given by the recursion:

$$D_p^m(\mathbf{K}) = \sum_{i=1}^{\min(m,p)} \binom{m}{i} (P_e)^i (1 - P_e)^{(m-i)} D_{p-i}^i(\mathbf{K}), \tag{6.17}$$

$$D_0^m(\mathbf{K}) = (1 - P_e)^m, \tag{6.18}$$

[3] For a failed transmission, we wait the SIFS time and the time needed to decode the ACK. After that time we can be sure the ACK is not received.

in which $\binom{m}{i}$ denotes the number of combinations to select i fragments out of m. The resulting probability to deliver a frame (JFR) is:

$$1 - JFR_p^m(\mathbf{K}) = \sum_{j=0}^{p} D_j^m(P_e, \mathbf{K}). \tag{6.19}$$

Time and energy requirements are respectively given by:

$$TXOP_p^m(\mathbf{K}) = [mT_{good}(\mathbf{K})] + [pT_{bad}(\mathbf{K})], \tag{6.20}$$

$$E_p^m(\mathbf{K}) = \sum_{j=0}^{p} D_j^m(\mathbf{K}) \times \{mE_{good}(\mathbf{K}) + jE_{bad}(\mathbf{K})\} + Z_p^m(\mathbf{K}) + H_p^m(\mathbf{K}). \tag{6.21}$$

The expected energy for a given configuration is the sum of the probabilities that the transmission will succeed after m good and j bad transmissions multiplied by the energy needed for good and bad transmissions. A second term $Z_p^m(\mathbf{K})$ should be added to denote the energy consumption for a failed job, hence when there are less than m good transmissions, and $(p+1)$ bad ones:

$$Z_p^m(\mathbf{K}) = JFR_p^m(\mathbf{K}) \times \left\{ E_{bad}(\mathbf{K}) + \sum_{j=0}^{m-1} D_{p+1}^j(\mathbf{K}) \times (jE_{good}(\mathbf{K}) + (P+1)E_{bad}(\mathbf{K})) \right\}. \tag{6.22}$$

The third term $H(\mathbf{K})$ denotes the cost that has to be added once every scheduling period. We will show later that this cost corresponds to a wake-up cost only and no reconfiguration cost should be taken into account, where we assume that the cost for each configuration $\mathbf{K}$ is constant.

We determine the Energy-Time-JFR trade-off as a function of the system state and number of retransmissions for each $\mathbf{K}$. This specifies the full profile for the system, and is determined only once during design or calibration time. These can then be combined into the 3D profiles of the system (Fig. 6.9(a)). This cost, resource and quality profile information is stored in each node's driver. We further prune the 3D profile to obtain per-flow Energy-TXOP curves. Configuration points that do not meet the target JFR^* are pruned. Next, the convex minorant is computed in both Energy and TXOP dimensions. The resulting two-dimensional Pareto-optimal trade-off curves are shown in Fig. 6.12 for the considered channel and frame size scenarios.

6.4.2 Anticipative Control in the 802.11 MAC Protocol

Based on the Energy and TXOP curves for each node, the scheduler in the AP can efficiently derive a near-optimal resource allocation at run-time using the greedy

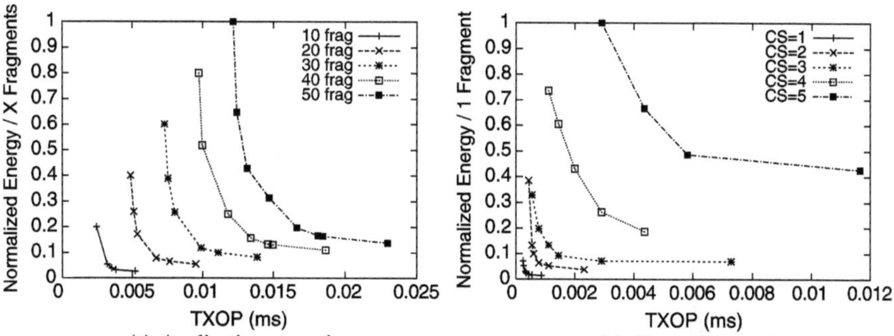

(a) Application scenarios. (b) Channel scenarios.

Fig. 6.12 The resulting Energy-TXOP Pareto-optimal trade-off curves to be combined at run-time to achieve the network optimum

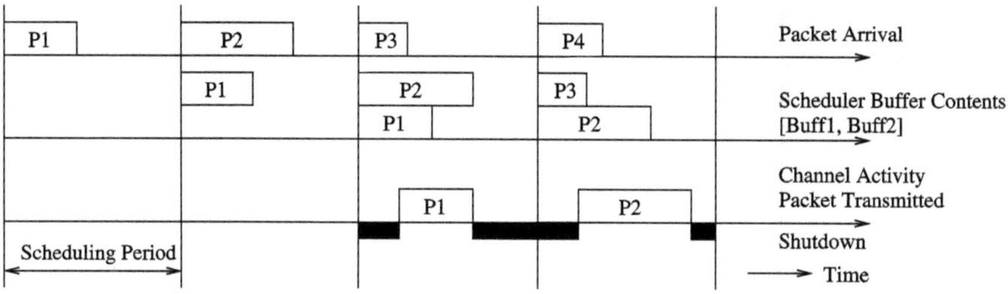

Fig. 6.13 MAC with two-frame buffering in the Scheduler Buffer to remove data dependencies and maximize sleep durations. By the third period of the single flow shown, frames 1 and 2 are buffered and frame 1 begins service. As the transmission duration of frame 2 is known at this time, the sleep duration between completion of frame 1 until the start of service of frame 2 is appended in the MAC header

scheme described in Sect. 6.3.2. The scheduler requires feedback on the state of each user and then communicates the decisions to the users. In order to instruct a node to sleep for a particular duration, the AP needs to know when to schedule the next packet. Waking a node earlier than the schedule instance will waste energy. Buffering just two frames informs the AP of the current and also the next traffic demand, allowing a timely scheduling and communication of the next period TXOP. In the ACK, the AP instructs the node to sleep until the time of the next TXOP meanwhile also communicates the required configuration. The AP now communicates with each node only at scheduling instances (Fig. 6.13). As the real-time packets are periodic, we eliminate by doing so all idle time between transmission instances. When a node joins a network, it sends its cost, resource and quality curves (stored in its driver) to the AP during the association phase. The AP then stores this and refers to it during each scheduling instance. By adding just three bytes in the MAC header for the current channel state and the two buffered frame sizes, each node updates the AP of its current requirement in every transmission. Protocols such as 802.11e provide support for this communication and therefore require only minor modifications. The AP then fetches the predetermined set for the current state from memory. In the case study, this corresponds to less than 3000 bits per state. A software-based QoS Module within the AP's network management layer maintains the list segments of all associated nodes and processes the current state of each node. At the beginning of each period, it executes the run-time phase of the control method and determines the configuration for each node during that period. The scheduling period requirement is determined by the rate at which the system state varies. Channel measurements show coherence times of 166 ms for stationary objects and moving scatterers [37]. Given a video frame rate of 30 ms, it is clear that this requires a scheduling period less than 30 ms. Since this timing requirement is rather low, a software module is sufficient. Alternatively, the QoS module can be integrated in a light-weight RTOS present in most embedded devices (as done in [89]). We expect the performance of the control method be lower for mobile networks with faster state dynamics, when it is difficult to feedback the system state timely. Faster adaptation schemes will be needed that integrate the adaptation module in hardware close to the physical layer [90].

6.5 Adapting to the Dynamic Context

We now show how a cognitive radio can save energy by optimally adapting to the varying context. More specifically, the smart radio knows how to adapt its flexibility across many layers to achieve the best configuration in terms of QoS, energy or resource use. The focus is on real-time streaming media applications with a reasonable target JFR^* set to 10^{-3}. In order to evaluate the relative performance of the cognitive control method, we consider four comparative transmission strategies:

1. **Smart:** This is the optimal scheme considering the energy trade-off between sleep and scaling, exploiting multi-user diversity. The node configurations are based on the profiles of Sect. 6.4.1.
2. **PHY-layer:** This scheme considers only physical layer scaling knobs. The Energy-TXOP profiles are set to scale maximally as no sleep mode is available.
3. **MAC-layer:** In this scheme only sleeping is possible by the energy-aware MAC-layer. The physical layer is fixed to the largest constellation and code rate, with maximum transmit power. This approach is used by commercial 802.11 devices [8]. However, the 2-frame buffering makes the proposed implementation more efficient as it eliminates all idle time between transmissions.
4. **Fixed:** Similar to MAC-layer scheme but the transceiver remains in idle after transmission. A basic scheme with no energy management and hence no dynamic range is exploited here.

The Energy-TXOP profiles are computed for the control dimensions considered in each scheme and used by the scheduling scheme implemented in the Network Simulator ns-2 [91]. This simulator has been extended with transceiver energy and performance models, and a slow fading channel model. Our simulation model implements the functions of the 802.11e with beaconing, polling, TXOP assignment, uplink, and downlink frame exchange, fragmentation, retransmission and variable super-frame

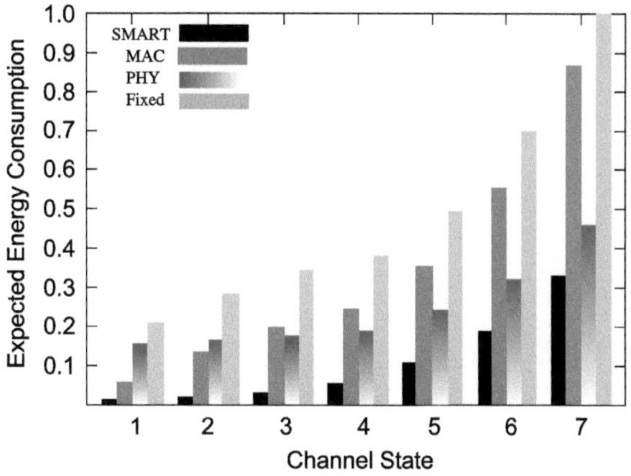

Fig. 6.14 Energy consumption across different channel states for 1 fragment

sizing. In all results, the total energy consumed by a node is averaged over a long duration to statistically capture the dynamics present in the scenario.

Consider the scenario where a single user has to deliver a unit frame every scheduling period. In Fig. 6.14, the energy consumption (normalized by the maximum energy consumed by Fixed) is plotted for the four schemes over different channel states. The anticipative adaptation method outperforms the other techniques in each state, since it takes advantage of the energy that can be saved by both sleeping and scaling. The energy needed to transmit a data fragment increases from best to worst channel state due to a combination of (a) the lower constellation needed to meet the PER (hence smaller sleep duration), (b) a higher required output power to account for the channel and (c) the increased cost of retransmissions. We observe, for example, that in the best channel state, the energy consumption is low for both the fixed and global approaches, and energy gains primarily result from sleeping. However, in the bad channel state the transmission energy is more important and scaling becomes more effective. The ratio of fixed to scalable energy varies as transceivers are designed differently. The take this into account during the calibration of system profiles. For a given platform, as users demand different levels of QoS, the proposed method jointly leverages the MAC and PHY to maximize the energy savings. A smart radio with control flexibility along many protocol layers, should know how to adapt to ensure an optimal user experience, energy use or resource use.

6.6 Conclusions

In this chapter it was shown how a flexible radio can be designed to meet the target QoS at minimal energy consumption for a set of anticipated scenarios. Flexible radios often have many different control knobs that can be tuned to save energy or boost performance. When many different knobs exist across many different layers, it becomes hard to anticipate the optimal configuration at design time. It was shown in this chapter how to achieve this for a 802.11a flexible radio. In the next section, it will be shown how a cognitive radio can be designed when the optimal configuration cannot be known perfectly at design time.

Chapter 7
Distributed Optimization of Local Area Networks

7.1 Introduction

In Chap. 5, we have demonstrated why the ISM bands can be considered the most complex wireless scene today. In these bands different heterogeneous networks and devices co-exist using only LBT etiquettes. The IEEE 802.11 standard is by far the most popular wireless standard for broadband access operating in the ISM bands. It is widely recognized that the throughput of IEEE 802.11 networks is dependent on all of the following parameters:

- Increasing the **transmission rate** will increase the number of packets that are sent during a time interval, but more packets can be lost at the receiver due to a lower interference tolerance.
- Lowering the **carrier sense threshold** protects a packet against interference, but decreases transmission opportunities by silencing the transmitter more often.
- Increasing the **transmission power** will decrease packet losses at the receiver, but will also reduce spatial reuse in the network.

In this chapter, we introduce a novel control algorithm, Spatial Learning, which learns the optimal operating point in this 3D design space at RT.

The model of IEEE 802.11 that we have presented in [46], can be used to investigate the relative importance of the different starvation mechanisms. However, the assumptions remain restricting for effective RT control. In this contribution, we have for instance assumed all terminals have the same power, rate and carrier sense threshold. It is unlikely that in an optimal setting, all terminals will have the same parameters. Indeed, terminals can be in different scenarios that demand for different settings. Hence, if we allow terminals to select their parameters individually the network throughput can further increase. Also, if we would need to ensure that all terminals have the same parameters, changes need to be communicated over the entire network [92]. The large overhead involved reduces throughput and possibly eliminates the adaptation benefit. Completely relying on DT modeling will hence not result in an effective RT control. Therefore, we have designed Spatial Learning relying on the framework presented in Chap. 3. The environment is split up into

S. Pollin et al., *Software Defined Radios*, Signals and Communication Technology,
DOI 10.1007/978-94-007-1278-2_7, © Springer Science+Business Media B.V. 2011

different scenarios and for each scenario a DT policy is built. However, to be detectable the scenarios used in this chapter cover a large amount of run-time situation and the optimal strategy for each run-time situation is not the same, as it's highly dependent on the situation of its neighbors. Therefore, learning becomes necessary to calibrate the DT policy for the observed scenario to optimally fit the experienced run-time situation.

Our novel algorithm exhibits the following desired properties for an IEEE 802.11 control algorithm [93]. It's a single-channel solution and maintains interoperability with other standards. Furthermore it is backward compatible with all IEEE 802.11 protocols as long as a terminal has a tunable power, rate and carrier sense threshold selection.

A lot of work has been done on optimizing these control parameters individually [94–96]. More recently, the optimization of two parameters at once has been investigated [97, 98]. To the best of our best knowledge, this chapter presents one of the first algorithms that tries to optimize all three parameters simultaneously.

This chapter is structured as follows. First, we present related work in Sect. 7.2. Our benchmark solution, Spatial Backoff, is explained in more detail in Sect. 7.2.2. We also explain the possibilities, remaining for future control algorithms. Our control algorithm, Spatial Learning, is discussed in Sect. 7.3. Simulation results are presented in Sect. 7.4. Finally, we present our concluding remarks in Sect. 7.5.

7.2 Existing Flexibility and Control Mechanisms

In this section we present some of the recent work on IEEE 802.11 optimization and present a general overview of multi-agent learning, i.e., when multiple terminals learn simultaneously.

7.2.1 Optimization of IEEE 802.11 Networks

Here, we present an overview of different algorithms that exploit the flexibility of transmission rate, carrier sense threshold or transmission power.

7.2.1.1 Transmission Rate

The main goal of rate adaptation algorithms is to determine the current channel state and select the optimal data rate accordingly.

One of the earliest rate adaptation algorithms is called Automatic Rate Fallback (ARF) [99]. Here, the rate is increased after S consecutive successful transmissions and dropped after F consecutive failures. ARF was improved by dynamically tuning S and F with Binary Exponential Backoff [100]. This avoids overprobing high rates and overutilization of low rates. Receiver-Based AutoRate (RBAR) monitors the Signal-to-Noise Ratio at the receiver, which is signaled to the transmitter using the RTS/CTS mechanism [101]. Opportunistic Auto Rate (OAR) tries to exploit the time when channel quality is high, by sending data packets back-to-back in these conditions [102]. In recent years, rate adaptation has focused on cross-layer techniques, loss-differentiation and more efficient feedback mechanisms [103–110].

7.2.1.2 Carrier Sense Threshold

Unlike the transmission rate, the flexibility of carrier sense threshold has been less explored. However, even recent papers show analytically the importance of an optimized carrier sense threshold for individual and system-wide throughput [46, 111, 112]. Here, we present some recent algorithms that try to optimize IEEE 802.11 networks by adapting the carrier sense threshold.

In [113], the authors show that the optimal choice of the CS threshold depends on node density and propose a distributed algorithm for local CS threshold optimization based on a local estimation of node density.

In [114], the authors use the carrier sense threshold to mitigate neighborhood starvation effects in ad hoc networks. This is done by detecting neighborhood starvation and increasing the carrier sense threshold in these situations.

In [115], the authors present an adaptation algorithm based on loss differentiation. Here, all stations measure the Packet-Error-Rate (PER). A linear adaptation algorithm, based on the highest PER in the network, is used to converge to an optimal value.

7.2.1.3 Transmit Power

Power control for IEEE 802.11 networks can target either throughput maximization, full network connectivity or energy efficiency. As the algorithm presented in this section targets throughput maximization, we will focus our overview on these algorithms.

In [116], the authors propose a control channel in order to inform other nodes about their interference tolerance. The RTS/CTS handshake is used to solve the asymmetric starvation problem.

In [94] it is recommended that links should scale down the transmission power until they see an increase of collisions. The goal is to decrease interference towards neighbors, without significant impact on the individual throughput.

In [95], the hidden and exposed node problems are balanced out using power control. The authors propose to decrease the transmission power of the link until only the interference area is covered. We have shown that this approach need not be optimal when considering neighbor location and neighborhood starvation [46].

7.2.1.4 Hybrid Control

Many algorithms exploit more than one parameter. Here, we present some of these hybrid control algorithms.

In [117] the authors present a power/rate adaptation algorithm that relies on the RTS/CTS exchange. They propose to extend the range of RTS and CTS packets to protect the CTS, DATA and ACK packets. This way, they try to remove all hidden terminals.

In [97] and [96], it is recommended to keep the product of transmit power and CS threshold constant in the network. In [97] this constant is found through Gibbs sampling, while in [96] the optimal constant is experimentally established when the network-wide PER is near-zero, which is optimal from a throughput perspective.

In [118], the authors investigate the effects of the 3 parameters on spatial reuse. They show that spatial reuse only depends on the ratio of transmit power to carrier sense threshold in the case that the achievable channel rate follows Shannon's law. Furthermore, they claim that tuning transmission power should be preferred over tuning the carrier sense threshold as terminals can then save energy additionally. Based on this observation, the authors present a power and data rate adaptation algorithm that uses the SINR as its input. Similar to [94], they suggest that terminals should use the minimum transmission power needed to sustain high data rates. Doing this, terminals achieve optimal throughput, while maximizing spatial reuse.

In [98] a spatial backoff contention resolution is presented. This is a heuristic algorithm that tunes both carrier sense threshold and rate and we will use it to benchmark the performance of our hybrid DT/RT optimization algorithm.

7.2.2 Benchmark Solution: Spatial Backoff

In this section, we introduce Spatial Backoff (SB) as our main reference [98]. This algorithm tunes both rate and carrier sense threshold to optimize individual throughput in IEEE 802.11 networks. We present an overview of the algorithm in Sect. 7.2.2.1 and present opportunities for further improvements of the algorithm in Sect. 7.2.2.2.

7.2.2.1 Algorithm Overview

Spatial Backoff assumes that the IEEE 802.11 MAC protocol has a set of discrete rates available. For each rate an associated carrier sense threshold, T_{CS}, is defined as follows:

$$T_{\mathrm{CS}}[i] = \frac{T_{\mathrm{Rx}}}{\mathrm{SINR}[i]}, \tag{7.1}$$

where T_{Rx} is the receiver sensitivity (no packets with a power below T_{Rx} can be received) and $\mathrm{SINR}[i]$ is the signal-to-interference-noise threshold for rate $R[i]$.

Only when the SINR at the receiver exceeds SINR$[i]$, packets can be received. It is assumed that when using a T_{CS} equal or lower than $T_{\mathrm{CS}}[i]$ to transmit at rate i, it is unlikely that the transmitter will overestimate the interference tolerance at the receiver. Hence, we are looking for a $T_{\mathrm{CS}} \leq T_{\mathrm{CS}}[i]$.

Transmitters start at the lowest rate, $R[0]$ and its associated $T_{\mathrm{CS}}[0]$. When a certain number S of consecutive transmissions succeed, the rate is increased to $R[i+1]$, unless it's already transmitting at the highest rate. Likewise, when a certain number F of consecutive failures is seen, transmitters try to compensate by first lowering their T_{CS}. This is done as long as the T_{CS} is higher than $T_{\mathrm{CS}}[i]$. When the current T_{CS} of the transmitter is equal to $T_{\mathrm{CS}}[i]$ and transmissions continue to fail, the rate is decreased to $R[i-1]$. When decreasing to a lower rate, T_{CS} is reset to its last known successful value for $R[i-1]$.

In order to stabilize on a certain point the number of consecutive transmissions needed to increase rate is increased $S[i]$, increases each time the transmitter falls back to rate i. A similar procedure is used for failures.

7.2.2.2 Opportunities

In this section, we shortly describe opportunities that exist to improve SB. In the next sections, we address these opportunities in the design of our control algorithm.

- **Improve monitors**
  SB uses a starvation detection mechanism to avoid a complete loss of transmission opportunities. This mechanism (no transmission during a certain period of time) is quite slow and only detects the most severe cases of starvation, when a transmitter is deprived of all access to the medium.
- **Increase the number of scenarios**
  SB only considers collisions. Collisions, however, are not always a good indicator for throughput as a node may increase its throughput by using decreasing its transmission rate, while increasing its carrier sense threshold. This can significantly increase the successful transmission opportunities.
- **Improve DT policy**
  The selection of $T_{\mathrm{CS}}[i]$ in SB is also quite defensive. Indeed, when the received power is high, the transmitter doesn't need to be as protective as when the received power is equal to T_{Rx}.
- **Improve RT policy**
  The proposed converging method can be quite slow, when the transmitter has a lot of neighbors and occasionally a packet fails through a coordinated collision [92].

7.2.3 Multi-Agent Learning

In this section we give a short overview of the current research on multi-agent and heuristic learning. Thorough reviews of the current state of research in these fields can be found in [119–122].

In [123], the authors describe what kind of engineering problems naturally lend themselves to multi-agent learning. They strongly support the development of multi-agent learning techniques starting from the application perspective. Also in [120], the authors present a critical view on the advancement of multi-agent learning techniques. They criticize the quest for convergence to equilibria, as these may be suboptimal (e.g., the Nash equilibrium in the prisoner's dilemma). They claim researchers should start from the problem and see how they can improve over state-of-the-art algorithms using these learning approaches.

In this chapter, we have tried to adopt this vision and have selected one of the simplest multi-agent learning techniques, independent q-learning, and see how we can use it in an engineering problem. Although convergence of independent q-learning is not guaranteed, it has been shown that it can outperform those Nash equilibria by finding mutually beneficial configurations [124].

Nash q-learning is an alternative that is known to converge to a Nash equilibrium. However, it requires to know the actions taken by all players. As the interference ranges for IEEE 802.11 are larger than the transmission ranges, this is not possible in our current scenario. Under these informational constraints, it has been proved in [125] that, today, no algorithm exists that can converge to the Nash equilibrium. This makes independent q-learning a viable candidate for the current problem.

In line with Chap. 3, we design DT procedures (heuristics) to speed up RT convergence of the algorithm. Here, the use of these DT procedures can also improve performance of the steady-state result. The use of heuristics improves the *blind* learning methods, through the addition of domain knowledge in the learner. As mentioned above, the authors of [120, 123] strongly support this.

7.3 Spatial Learning: Distributed Optimization of IEEE 802.11 Networks

In this section, we explain our control algorithm for distributed optimization of IEEE 802.11 networks, resulting from an instantiation of the smart radio framework presented in Chap. 3. First, we introduce the instantiated framework in Sect. 7.3.1 and make the link with Chap. 3. Afterwards, we introduce the control dimensions in Sect. 7.3.2. Scenario identification is discussed in Sect. 7.3.3. Afterwards, we introduce our DT procedures in Sect. 7.3.4. The learning engine to calibrate these DT procedures is discussed in Sect. 7.3.5. In Sect. 7.3.6, we explain how the DT procedures and the learning engine fit together. Finally, we end this section with a discussion on implementation details in Sect. 7.3.7.

7.3.1 The General Framework

In this section, we introduce the general framework of our new control algorithm, Spatial Learning (see Fig. 7.1). Similar to the CR cycle in [16], the learner interprets

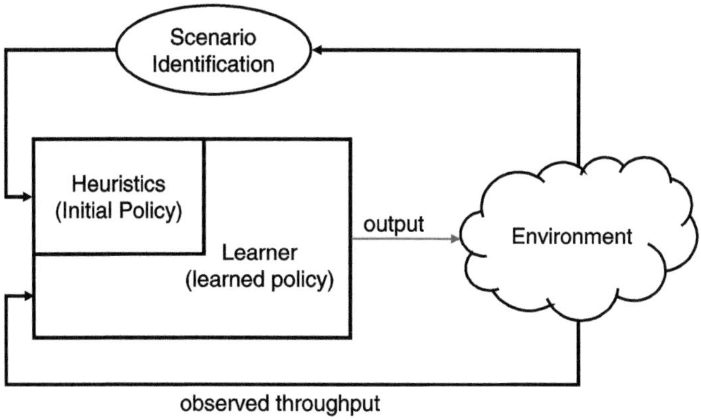

Fig. 7.1 General framework

how the environment reacts to the selected actions and adapts his actions accordingly. As mentioned by Simon Haykin in his seminal paper: "The CR is therefore, by necessity, an example of a feedback communication system". Hence, feedback is an essential part of the CR and, hence, of Spatial Learning.

In our current IEEE 802.11 scenario, the feedback from the environment is the observed throughput. As mentioned above, this limited information assumption excludes the guaranteed convergence to a Nash Equilibrium, since the exact actions taken by all players cannot be known [125]. As shown in Fig. 7.1, the learner is assisted by heuristics, i.e., DT procedures. Heuristic recommendations for the learner are based on a scenario identification technique, that relies on local observations. This in effect also increases the feedback information quality from the environment, which speeds up convergence, while increasing the performance during convergence.

Below, we show how we instantiated the general framework in this case study chapter. We also refer to the specific sections, where our selections are discussed in more detail. As stated in Sect. 7.1, the considered control dimensions in this case study chapter are the transmission power, transmission rate and carrier sense threshold. In Sect. 7.3.2, we introduce the possible knob settings for each control dimension in the discrete state space. As defined in Chap. 3, actions are the relative change in knob settings. In this case study chapter, the actions are to increase or decrease each knob if possible. When decreasing the rate, the carrier sense threshold is also reset. Further details can be found in Sects. 7.3.4 and 7.3.5. The system scenarios we consider in this instantiation are the hidden terminal starvation, neighborhood starvation and asymmetric starvation and starvation-free scenario. The exposed terminal problem cannot be monitored by local observation of the IEEE 802.11 terminal so it cannot be considered as a system scenario. Instead, it is clustered in the starvation-free scenario. Further details are presented in Sect. 7.3.3. We detect the different system scenario using simple energy detection and collision detection. The exact procedure of scenario identification, based on these detection mechanism is described in Sect. 7.3.3. To calibrate the DT policies, we use the theoretical saturated throughput as a feedback channel. The theoretical throughput is used, as opposed to the actual observed throughput, to reduce the noise generated by the backoff selection mechanism. The calculation of the theoretical throughput requires the knowledge of the busy, idle, successful and collision probabilities. The feedback channel is further discussed in Sect. 7.3.5.

7.3.2 The Control Dimensions

Here we describe in more detail the set of possible knob settings for our algorithm. This state space is selected at design time by the operator. We restrict it to allow faster convergence of the algorithm. Increasing the number of possible configurations decreases convergence speed, but possibly allows more optimal results.

7.3.2.1 Rate

Due to the underlying modulation schemes, the rate is the most coarse-grained parameter and hence does not need any design-time restriction. We use all rates the platform supports and use these to bound the more flexible parameters: carrier sense thresholds and power.

7.3.2.2 Transmission Power

As a rule-of-thumb, we select the number of transmission powers equal to the number of available rates and distribute these uniformly over the interval $\left[\frac{\max(P_t)}{n_R}, \max(P_t)\right]$, where n_R is the number of available rates and $\max(P_t)$ the maximum available transmission power.

7.3.2.3 Carrier Sense Threshold

Here, we describe how we select the definition of $T_{CS}[i]$, which is the most flexible parameter from an implementation point of view. Rather then relying on the same states for each transmitter as done in [98], we allow the transmitter to select its $T_{CS}[i]$ based on the path loss:

$$T_{CS}[i] = \left(\frac{P_r}{P_t}\right)^k \left(\frac{P_r}{\text{SINR}[i]} - \sigma_0^2\right), \tag{7.2}$$

where P_t is the transmit power, P_r is the received power and σ_0^2 a model for the noise power at the receiver. The operator can choose the parameter k from the interval $[-1, 1]$. The aggressive point, $k = -1$, assumes that the interference power suffers from the same path loss as the link between transmitter and receiver. The defensive

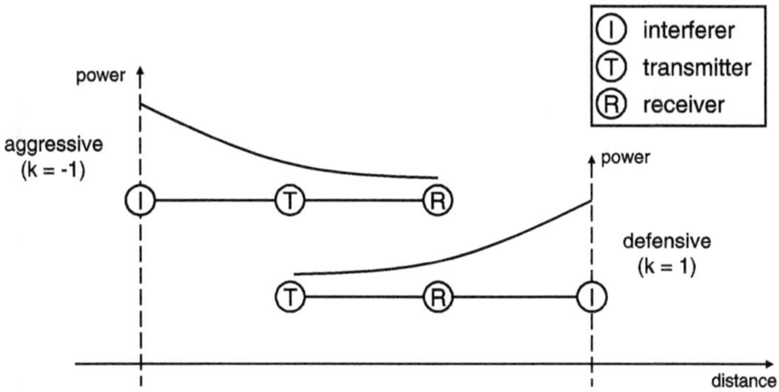

Fig. 7.2 The network operator can select its k value from the interval $[-1, 1]$. When the interferer is assumed to be closest to the transmitter, an aggressive strategy can be used ($k = -1$). When the interferer is assumed to be closest to the receiver, a defensive strategy needs to be employed. This defines the selection of the carrier sense threshold according to (7.2)

point, $k = 1$, assumes that the interference power, when measured at the transmitter, suffers an extra path loss from receiver to transmitter (see Fig. 7.2). Throughout the remainder of this chapter, we will use a neutral setting ($k = 0$).

This CS setting implies that the transmitter needs to know the value of P_r. A commonly accepted way is to piggyback the received power estimated at the receiver in the acknowledgment. As mentioned in [126], this has already been integrated in current IEEE 802.11 commercial products (e.g., the Lucent ORiNOCO AP-1000 card). Even when it would be difficult to estimate the received power, SL will compensate by selecting appropriate states within the state space.

We denote this improved state space as P_r*-space*. The state space described in Sect. 7.2.2.1, will be referred to as T_{Rx}*-space*.

7.3.3 System Scenarios

In order to design the DT procedures (i.e., heuristic recommendations), we must first find out which is the current scenario. We use scenario identification techniques, quite similar to those presented in [92]. The scenarios we can identify are the hidden terminal starvation, asymmetric starvation and neighborhood starvation and the starvation-free scenario. Below, we describe how we can identify these different scenarios.

- Hidden terminal starvation is the easiest to detect, as only an inspection of p_c is needed. Indeed, it is signaled through a high collision probability.
- Asymmetric starvation is detected as follows. We keep a timer from the beginning of transmission until we can decrement our backoff timer. If a successful transmission lasts longer than expected (the expected duration of a collision or a successful transmission can be easily calculated), a neighboring node started

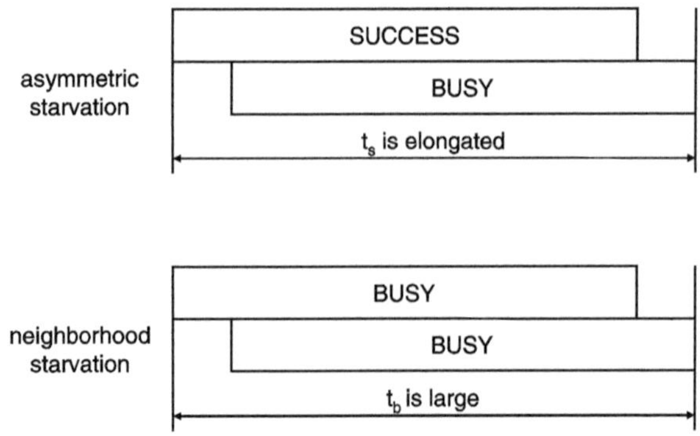

Fig. 7.3 Different types of starvation mechanism and the way they are detected

transmitting during our transmission. This means that this node was not listening to our transmission, but we are listening to his transmission. Hence, this is a clear case of asymmetric starvation (see Fig. 7.3).

- Neighborhood starvation is detected by long busy times (see Fig. 7.3). Using this detection mechanism, we cannot distinguish between slow neighbors and neighborhood starvation. However, as the recommendation for this situation is the same (you want to increase T_{CS}), this does not really matter.
- The starvation-free scenario occurs when the above three scenarios are not identified.

7.3.4 Design-Time Procedures

The optimal operating point cannot be known at DT. However, for each scenario we can identify possible escape procedures that may lead to better performance. We rely on the learner to identify at RT the best actions among those possible. However, we can already at DT devise procedures, also called heuristic recommendations hereafter, that are used as a seed for the learning engine.

In Fig. 7.4, we see the system scenario diagram. As the system scenario diagram is based on the scenario identification techniques presented in Sect. 7.3.3, it differs from the state space presented in Sect. 7.3.2. The heuristic recommendations are also shown and explained below.

Hidden Terminal Starvation

The hidden terminal starvation occurs when a large amount of packets fail (see Sect. 7.3.3). This implies that the SINR at the receiver is too low for successful packet transmission. Hence, the heuristic recommendation would be to either to increase the SINR at the receiver or to lower the required SINR for successful packet reception. This can be done by:

- increasing the transmission power: this increases the received power and, hence, increases the SINR at the receiver,

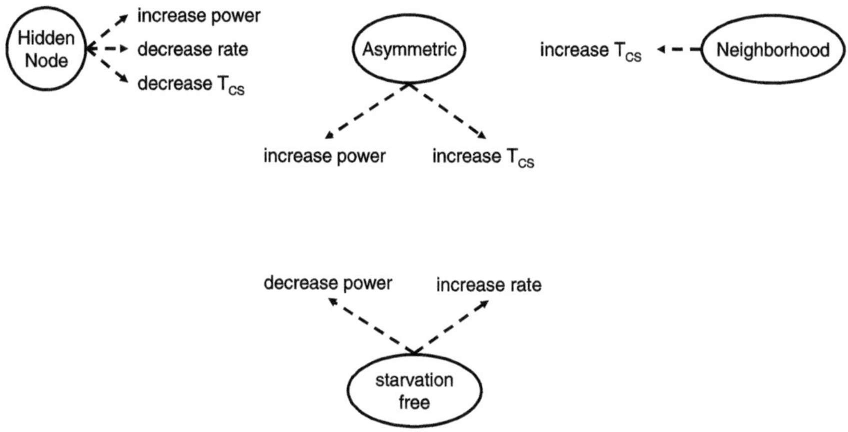

Fig. 7.4 The system scenarios along with their heuristic recommendations

- decreasing the carrier sense threshold: this lowers the interference at the receiver and, hence, increases the SINR at the receiver,
- decreasing the transmission rate: this lowers the required SINR for successful packet reception at the receiver.

Asymmetric Starvation

The asymmetric starvation problem occurs when terminal A is backing off for terminal B. However, terminal B is not backing off for terminal A. Thus, terminal A can try to remove this scenario by:

- increasing the transmission power: terminal A wants to make himself heard by increasing the received power at terminal B,
- increasing the carrier sense threshold: terminal A wants to remove the asymmetric starvation scenario by not listening to terminal B anymore.

Neighborhood Starvation

We have modeled neighborhood starvation in [46]. The only local action a terminal can take to remove this scenario is to:

- increase the carrier sense threshold.

Starvation-Free Scenario

Also for the most ideal scenario, we can give heuristic recommendations:

- increase the transmission rate: the terminal tries to see whether a higher transmission rate can be sustained,
- decrease the transmission power: the terminal tries to see whether the current transmission rate can be sustained at a lower transmission power. This decreases the interference towards other terminals, allowing them to increase their transmission rate.

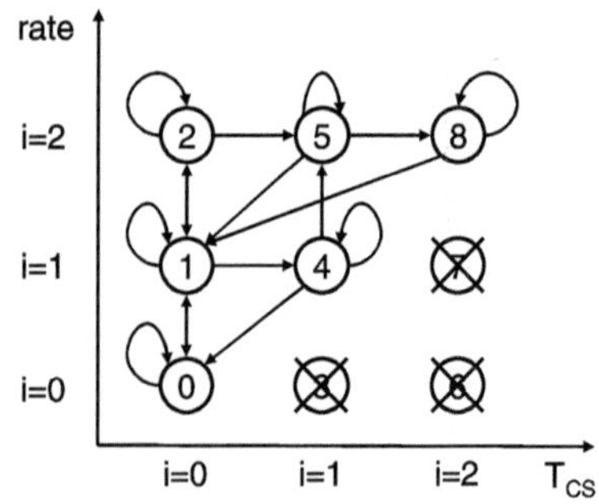

Fig. 7.5 In each state the transmitter can decide (if possible) to stay, increase rate, decrease T_{CS} or decrease rate. When decreasing the rate, T_{CS} is reset to $T_{\mathrm{CS}}[0]$. The states where T_{CS} is smaller than $T_{\mathrm{CS}}[i]$ have been pruned at DT

7.3.5 The Learning Engine

To find the optimal operating point, we propose to use a learning based approach to refine the heuristic recommendations. The proposed learning scheme is Q-learning, which does not need a model of the environment and can be used on line [67].

The Q-learning algorithm continuously estimates the values of state-action pairs. In the current problem, an action is the transition from the current state (s_c) to a new state (s_n). The value $Q(s_c, s_n)$ is defined as the expected discounted sum of future payoffs obtained by going to the new state and following an optimal policy thereafter. Once these values have been learned, the optimal action from any state is the one with the highest Q-value. The standard procedure for Q-learning is as follows. All Q-values are initialized to 0. During the exploration of the state space, the Q-values are updated as follows:

$$Q(s_c, s_n) \leftarrow (1-\alpha)Q(s_c, s_n) + \alpha\big[(1-\gamma)r + \gamma \max_{s\in\mathcal{A}(s_n)} Q(s_n, s)\big], \tag{7.3}$$

where α is the forget factor and γ is the learning parameter.

To implement Q-learning we first need to define the actions within the state space. For simplicity reasons, we discuss first the actions without power flexibility. The possible actions for an example with 3 available rates can be seen in Fig. 7.5. In each state the transmitter can decide (if possible) to keep the current configuration, to increase the rate, to decrease the carrier sense threshold or to decrease the rate. When decreasing the rate, the transmitter is forced to reset its T_{CS} to $T_{\mathrm{CS}}[0]$. The states where T_{CS} is smaller than $T_{\mathrm{CS}}[i]$ have been pruned as explained in Sect. 7.3.2.3.

The reward r is defined as the throughput increase for going to state s_n:

$$r = S(s_n) - S(s_c), \tag{7.4}$$

where $S(s)$ is the throughput seen in state s.

The Q-learning algorithm updates the estimate for each action, but in fact does not specify what actions should be taken. It allows arbitrary experimentation while at the same time preserving the current best estimate of the states' values. For now, we will use the following instantiation of simulated annealing. The nodes explore using a soft-max policy, where each node selects an action with a probability given by the Boltzmann distribution [127]:

$$p_Q(s_c, s_n) = \frac{e^{\frac{Q(s_c, s_n)}{T}}}{\sum_{\forall s \in \mathcal{A}(s_c)} e^{\frac{Q(s_c, s)}{T}}}, \tag{7.5}$$

where T is the temperature that controls the amount of exploration. For higher values of T the actions are equiprobable. By annealing the algorithm (cooling it down) the policy becomes more and more greedy. We use the following annealing scheme, denoting θ the annealing factor:

$$T_{k+1} \leftarrow \theta T_k. \tag{7.6}$$

To further improve network-wide throughput and fairness, we allow transmitters to tune their power. It has been established that the best response is to always send at a higher power. This will lead to the Nash equilibrium, where all terminals are using the maximum power. Hence, we need to give nodes a small incentive to scale down the power. We do this by introducing a cost for using higher powers:

$$r^{(p)}(s_n, s_c) = \overline{\rho}(i_n) S(s_n) - \overline{\rho}(i_c) S(s_c), \tag{7.7}$$

where i is the power index ($i = 0$ refers to the lowest power). The reward with power rewards is denoted as $r^{(p)}$. The reward factors are defined as follows:

$$\overline{\rho}(i) = \rho^i: \quad i \in [0, n_p - 1], \tag{7.8}$$

where ρ is an element of $(0, 1]$ and n_p is the number of available transmission powers.

With a high ρ, nodes will scale down their power, until they see a drop in throughput. This is somewhat similar to the power control mechanism described in [94]. With a lower ρ, they will even accept a throughput reduction.

Similar to the heuristic recommendation for starvation-free scenarios (see Sect. 7.3.4), we allow links with a *good* channel to scale down their power without dropping their throughput. As a result interference levels drop for the surrounding links. These links may now be able to send at a higher rate, which improves network-wide throughput and fairness.

7.3.6 Seeding the Learning Engine with the DT Procedures

For each (combination of) scenario(s), we have defined heuristic recommendations in Sect. 7.3.4. For instance, when a node is dealing with asymmetric starvation, it makes sense to either increase the power or decrease the carrier sense threshold in order to alleviate this situation. At DT, we however do not know which of the two actions is better.

Hence, we need to incorporate the heuristic recommendations in the Q-learning mechanism, described above. The idea is that the heuristic recommendations are followed during the exploration phase and that when the temperature cools down, the Q-learning takes over from the heuristic. This can be achieved by redefining (7.5) as follows:

$$p_{HQ}(s_c, s_n) = \frac{p_H(s_c, s_n) e^{\frac{Q(s_c, s_n)}{T}}}{\sum_{\forall s \in \mathcal{A}(s_c)} (p_H(s_c, s) e^{\frac{Q(s_c, s)}{T}})}. \tag{7.9}$$

An equiprobable distribution for p_H turns (7.9) back into regular Q-learning. The heuristic recommendations limit exploration for the Q-learning by avoiding that the Q-learning algorithm visits dominated states. This speeds up convergence and increases performance during convergence, as explained above.

We introduce the concept of heuristic bias. The heuristic bias, h_b, defines how much focus we put on the heuristic recommendations. The heuristic probabilities are formed by weighing the number of recommended actions with h_b and the action to be avoided with $\frac{1}{h_b}$. The heuristic probabilities are then calculated as follows:

$$w_H(a^{(+)}) = h_b, \tag{7.10}$$

$$w_H(a^{(0)}) = 1, \tag{7.11}$$

$$w_H(a^{(-)}) = \frac{1}{h_b}, \tag{7.12}$$

$$p_H(a) = \frac{w_H(a)}{\sum_{i=1}^{n_a(s_c)} w_H(i)}, \tag{7.13}$$

where w_H is the heuristic weight, $a^{(+)}$ is a recommended action, $a^{(0)}$ is a neutral action and $a^{(-)}$ is an action to be avoided. The number of actions in a certain state is denoted as $n_a(s_c)$.

7.3.7 Implementation in the IEEE 802.11 MAC Protocol

We have defined reward and actions in Sect. 7.3.5. Here, we detail the implementation of the algorithm in the IEEE 802.11 MAC protocol. The basic concept is to stay in one state for a certain time t_u. After this timer has passed, the Q-values are updated and a new state is selected according to (7.5).

7.3.7.1 The Reward

As we want to track the throughput as our reward (see Eqs. 7.4 and 7.7), a logical choice would be to count the number of acknowledgments received during t_u. However, the number of packets is highly dependent of the actual backoff values used during the observation time. This noise may cause the learner to misinterpret the rewards.

7.3.7.2 Theoretical Throughput Estimation

To remove the noise generated by the correlation between backoff selection and observed throughput, we use the theoretical saturation throughput, $S^{\star}$, of IEEE 802.11 given by (13) in [128], rather than the actual observed throughput ($r(s) = S^{\star}(s)$). This theoretical throughput is calculated based on 5 parameters: the collision probability p_c, the busy probability p_b, the length of a collision slot t_c, the length of a busy slot t_b and the length of a successful slot t_s.

By using transmissions as probes for p_c, t_s and t_c and observing p_b and t_b when the transmitter is not sending, we can calculate the theoretical throughput. This theoretical throughput is no longer dependent on the actual instantiations of the backoff value.

7.3.7.3 Observation Reuse

As the observed throughput is still dependent on the backoff values and actions of other transmitters in the neighborhood, not all noise is removed. This may cause the learner to still misinterpret the rewards as shown in Fig. 7.6. As the transmitter has no way of knowing the actions of other players (see Sect. 7.2), we try to accommodate this by averaging out these variations by observing the following relations:

$$p_i = f(P_t, T_{\text{CS}}), \tag{7.14}$$

$$t_b = f(P_t, T_{\text{CS}}), \tag{7.15}$$

$$p_c = f(R, P_t, T_{\text{CS}}), \tag{7.16}$$

$$t_c = f(R, P_t, T_{\text{CS}}), \tag{7.17}$$

$$t_s = f(R, P_t, T_{\text{CS}}). \tag{7.18}$$

We see that p_i and t_b are not rate-dependent. This means that the observation of p_i and t_b in one state is also valid for all states that have the same carrier sense threshold and transmission power. Hence, we are virtually checking these states and we can update also the observations in these states.

We can also see that p_c is dependent on all three parameters. However, if we enable SINR-measurements at the receiver as done in [101], we can also virtually update p_c for the states that have higher rates, but equal transmission power and carrier sense thresholds. In each acknowledgment the receiver piggybacks the SINR it observed during reception of the packet. Using this SINR, we can find out if a higher-rate packet would have succeeded or not. Obviously, each failed attempt would also have failed if it was transmitted at a higher rate. It is not possible to update the p_c observation for lower rates, because the packet wasn't received and, hence, no acknowledgment was sent. This means we have no way of finding out whether a failed transmission would have been successful at the lower rates.

To further reduce the variance on the observations, we use ARMA filters to keep track on all the individual parameters.

2 \ 1	L	R
U	(6,8)	(5,8)
D	(6,10)	(5,0)

(a) The payoff matrix of an illustrative example.

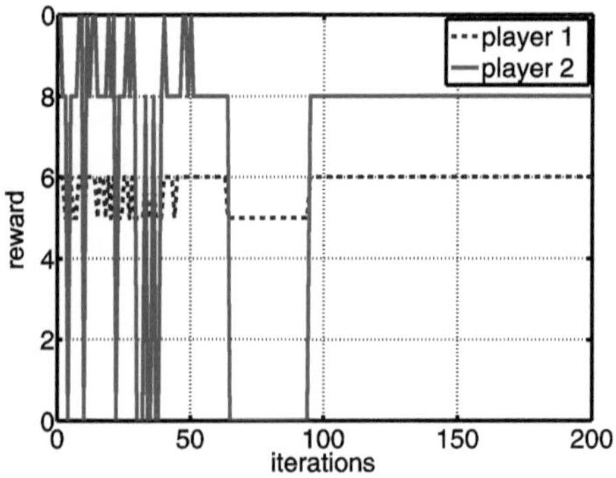

(b) The rewards received by both players in time.

Fig. 7.6 Multi-agent q-learning using simulated annealing is not guaranteed to converge to a Nash equilibrium. In this example, the Nash equilibrium is (L, D). However, during initial exploration, when all the actions are equiprobable, player 2 may decide that U is the better choice as it has an average profit of 8 and D only yields a profit of 5. As it is unaware of the actions taken by player 1, player 2 fails to notice that player 1 settles on L. Hence, player 2 goes for the *safe* option where it cannot be hurt by the exploration of player 1

Table 7.1 Spatial learning: simulation parameters

(Rate [Mbps], SINR [dB])	Power [mW]
(9, 7.78)	250
(18, 10.79)	125
(36, 18.80)	66.25
(54, 24.56)	33.125

7.4 Assessing the Gains

We evaluated the performance of the proposed algorithm through an extensive simulation study with the network simulator ns-2.29 [129].

All simulations are done using the simple path loss model of ns-2.29 without shadowing. As can be seen in Table 7.1, we assumed terminals are capable of transmitting at four discrete rates to make a fair comparison with [98]. We also used four power levels. For parameters not present in Table 7.1 or Table 7.2 we used the default values in ns-2.29.

First, we present an illustrative example that demonstrates the benefits of the different contributions. The scenario is presented in Fig. 7.7(a). In Fig. 7.7(b), we can see that the T_{CS} based on T_{Rx} is too defensive. Link 1 can increase its throughput with a smarter state space selection. This, however, has a negative impact on link 2. When using SL without power flexibility, we can quickly converge. When we allow link 1 to decrease its power, interference for link 2 reduces and it is able to sustain a higher rate.

Table 7.2 The parameters of the IEEE 802.11 MAC protocol

$\underline{W}$	15	$\overline{W}$	1023
SIFS	16 μs	σ	9 μs
T_{preamble}	16 μs	l_{ACK}	14 bytes
l_{H}	28 bytes		

As illustrated in Fig. 7.8, we assumed that 802.11 Access Points form a hexagonal grid. The User Equipments (UEs) are distributed according to a spatial Poisson point process with a density $\delta = 25\ \frac{\text{nodes}}{\text{km}^2}$. UEs are sending uplink traffic using the DCF mode. The results for a complete ad-hoc network are similar, but we believe this topology is a better representation of 802.11 networks today.

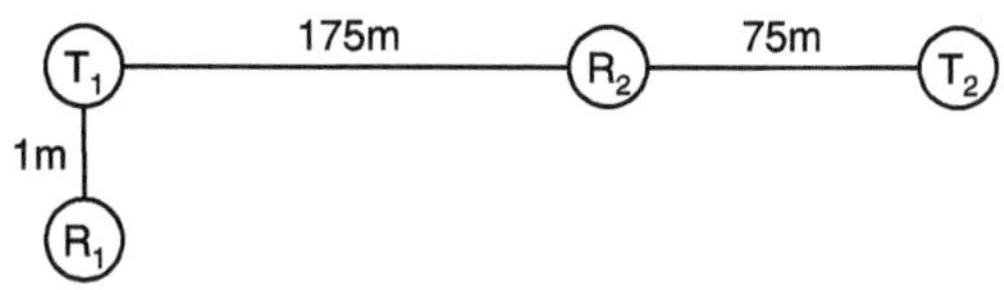

(a) An illustrative example to show the benefits of each contribution.

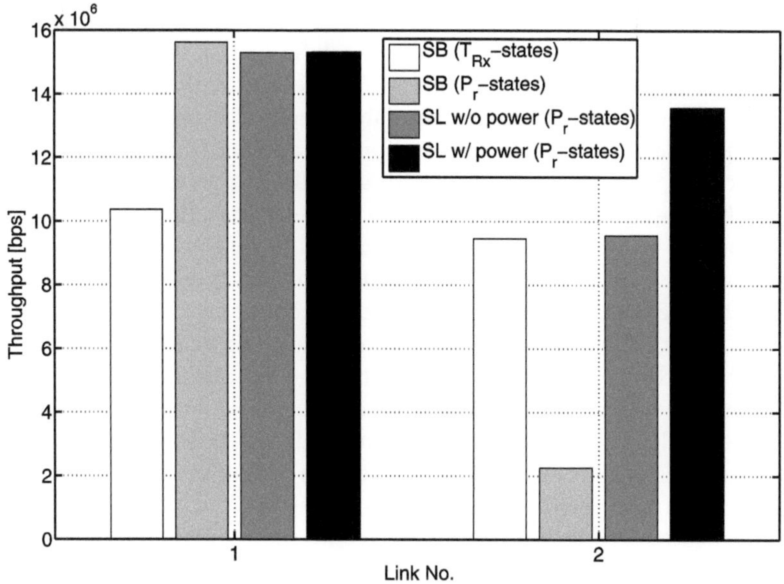

(b) Throughput during the 10th second of simulation of the scenario shown in Fig. 7.7(a).

Fig. 7.7 By allowing link 1 to use less defensive T_{CS}, it can increase its throughput. This causes a throughput drop for link 2. However, as link 1 reduces its power, while sustaining its throughput, link 2 can support a higher rate due to a decreased interference level

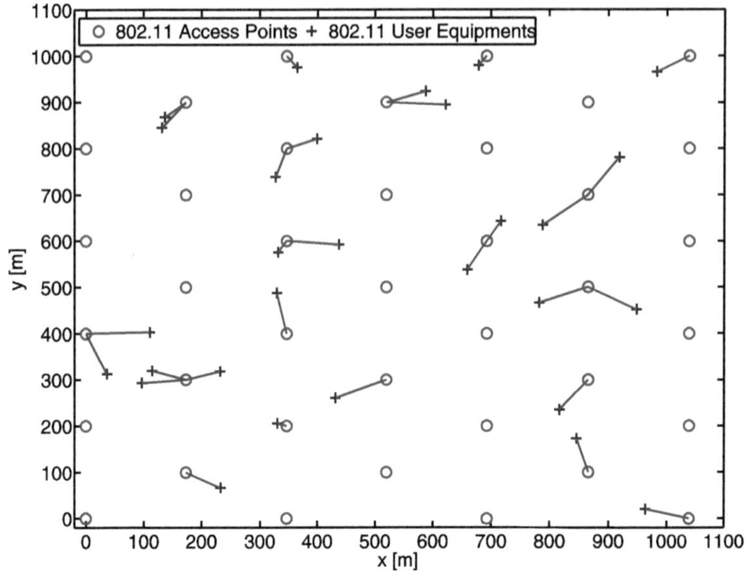

Fig. 7.8 A centralized topology: 802.11 Access Points form a hexagonal grid. 802.11 User Equipments are distributed randomly according to a spatial Poisson process. The UEs are associated with the nearest access point

In Fig. 7.9, we present a comparison between SB and SL. Results are averaged over 100 seeds, using the topology of Fig. 7.8. We can see that our new T_{CS} definition, based on P_r, significantly increases throughput. The definition based on T_{Rx} does not provide a lot of differentiation among the different states. This can be seen by the fact that SL cannot improve over SB using these states and by the fact that it reduces the transmission power significantly for these states. However the introduction of the new states also causes more unfairness in the network as terminals are less inclined to listen to each other and hidden terminals become more prominent. By allowing the nodes to scale down power, SL compensates for this and reaches the same levels of fairness[1] as the T_{Rx}-states. Most importantly, SL is shown to outperform SB by 33% in throughput using the P_r-based states.

The impact of heuristics on the convergence of SL is illustrated in Fig. 7.10. Again, the topology of Fig. 7.8 is used and averaged over 100 seeds. It can be clearly seen that the use of domain knowledge in the form of heuristics results in a significant increase in convergence speed.

To demonstrate the interoperability of SL, we show the performance of SL in a legacy 802.11 network in Fig. 7.11. As most of the current commercial 802.11 cards support some kind of ARF, we have implemented this for the legacy 802.11

[1] We use Jain's Fairness Index as an indicator for fairness. This index is calculated as follows: $f = \frac{(\sum_{i=1}^{n} S_i)^2}{n \sum_{i=1}^{n} S_i^2}$ [40].

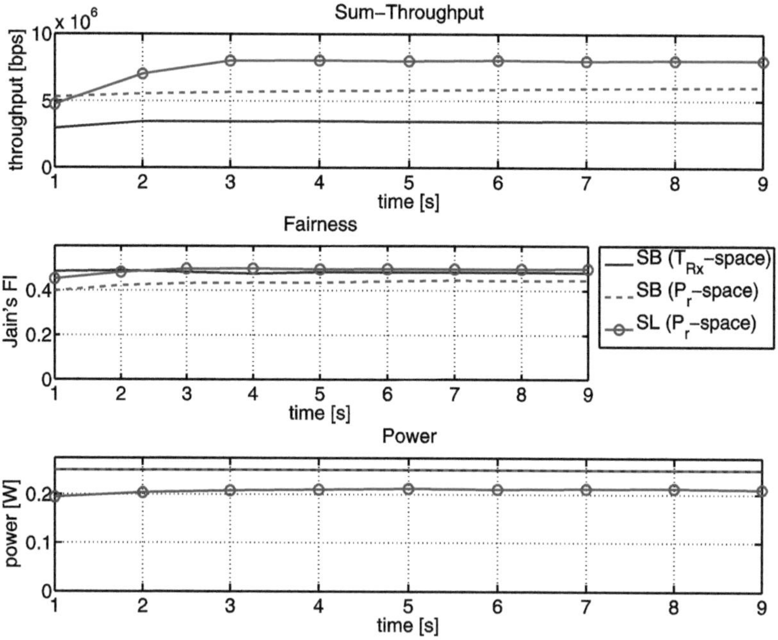

Fig. 7.9 Spatial Learning is shown to outperform Spatial Backoff by 33% in throughput using the P_r-based states. It also reaches a better fairness index and lower power than SB

Fig. 7.10 We use the heuristic recommendations to speed up convergence. The addition of heuristics to Q-learning also allows to converge to a better steady-state solution. When h_b is equal to 1, heuristics are not considered, as can be seen in (7.10)–(7.13)

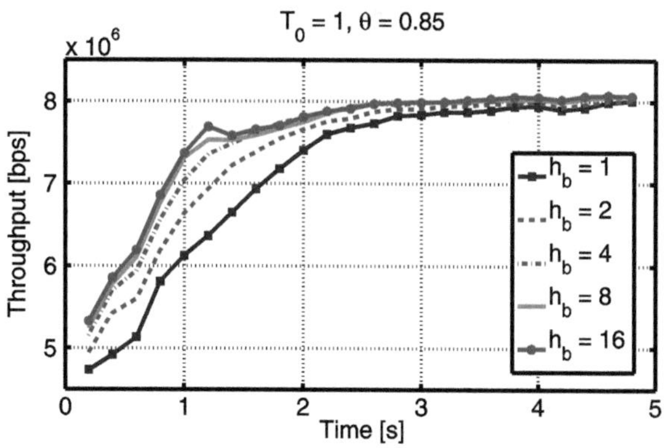

terminals [99]. We can see that SL performs better among legacy 802.11 terminals as these cannot optimize their own transmissions fully, generating opportunities for the SL terminals. However, adding more SL terminals doesn't reduce the average throughput of the SL terminals that much. This makes for a compelling business case, where the first adopters are rewarded the most.

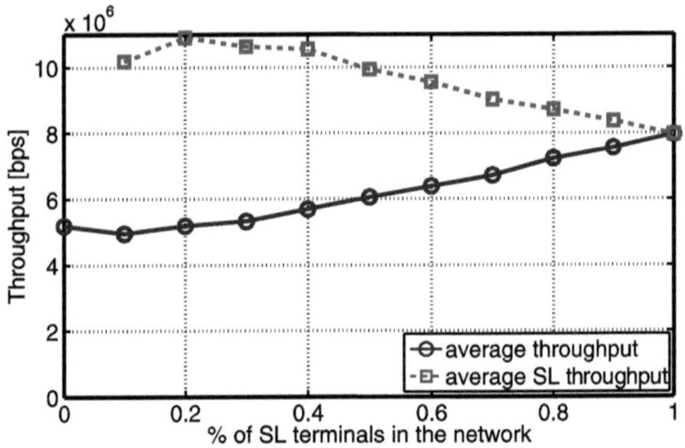

Fig. 7.11 In a legacy network, an SL terminal will perform significantly above average. With the introduction of more SL terminals, the average throughput of SL terminals begins to decrease as less terminals can now be exploited. However, a full SL network still outperforms a full legacy network

7.5 Conclusions

In this chapter we have presented a throughput model for large 802.11 networks. This illustrates that the DT/RT method can handle very complex environments for CRs. The proposed model is shown to match simulation results very accurately, and hence gives the correct insight in the relative importance of the different starvation sources on the expected throughput of large networks.

This model is however too rigid for RT control of IEEE 802.11 terminals. Hence, we have presented a hybrid DT/RT control algorithm for tuning power, rate and carrier sense threshold. It is shown to outperform the reference algorithm Spatial Backoff in network-wide throughput and power consumption. We believe distributed optimization is a key topic in wireless communications as it is able to cut down infrastructure and management cost. This is key for CR.

IEEE 802.11 has been selected as the supporting standard, since it is the most prominent wireless standard that enables ad-hoc communication. In the future, we expect a synergy between purely distributed optimization based on local information and distributed optimization based on information exchange between the terminals.

Chapter 8
Close

In this final chapter of this book, we first summarize its major conclusions. We introduced design and operation approaches towards Smart(er) and Cognitive Software Defined Radios, and showed its application in particular stringent cases. Consequently and finally, we end this chapter and book with some closing remarks. Many open challenges still remain to realize the *Anything, Anytime, Anywhere* aspiration, which definitely continues to be a moving target. Technological innovation is needed to keep up the pace with increasing capacity requirements. Moreover, regulations, economical and even social conditions have to be considered in order to ultimately serve the users optimally.

8.1 "Good Enough" Is "Close Enough to Optimal"

Wireless services have attracted an impressively growing interest over the past decades. This trend is expected to continue as ever more users want to exchange ever more and richer information. Despite great progress in physical layer techniques, we are however running out of wireless capacity. Novel solutions, both technological and regulatory, are therefore needed to enable a sustainable growth in wireless communication. This book focuses on opportunities opened up by the progress towards smart(er) and cognitive radios. In Sect. 8.2 of this chapter we draw the attention to complementary tactics which we believe will also be vital to keep the communication going. The smart(er) and cognitive operation of radios presented in this book, exploit the possibilities offered by recent and emerging standards which enable a more dynamic spectrum access (see Chap. 2). Consequently, radios that exploit these new opportunities to adapt their behavior to the actual circumstances (both in terms of service requirements and communication scenario) can achieve great performance increase. We thereby adapt the taxonomy illustrated in Fig. 8.1. While cognitive radios can especially bring added value when they are built upon flexible hardware (SDRs, reconfigurable radios), this is not the only form of flexibility to be exploited in the wireless environment. Therefore, the case studies discussed in this book do not impose flexible hardware as a prerequisite.

S. Pollin et al., *Software Defined Radios*, Signals and Communication Technology, DOI 10.1007/978-94-007-1278-2_8, © Springer Science+Business Media B.V. 2011

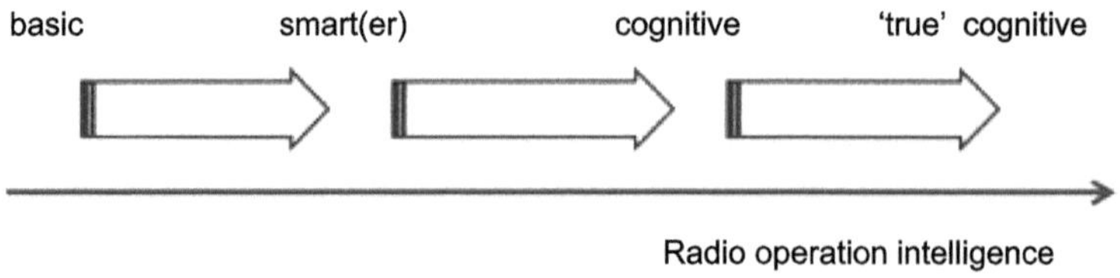

Fig. 8.1 Smart(er) to cognitive radio operation

We want to acknowledge to the readers that the boundaries between these classes are not easy to draw. While a crystal clear separation would be desired, it is our opinion that the move towards increased radio intelligence will be gradual, rather than step-wise and that this should considered good news. When J. Mitola launched the cognitive radio concept [16], many questions arose if and when this disruptive idea could be implemented in practical systems within regulatory constraints. A gradual and smooth transition is however feasible, as we hope to have shown in this book. "Good enough" is "close enough to optimal" is thereby the main message we want to convey when concluding this book. When searching for solutions for smart(er) and cognitive operation of radios, we follow the same principle. Certainly in a dynamic environment, such as the wireless scene, the search for ultimate optimal operation may take more energy than could be paid off by the actual gains. Moreover, by the time this optimum is found the conditions may have changed already. We have proposed a generic control strategy based on a two phased approach. In a preparation phase, at design time, a representative set of scenarios are analyzed and corresponding sets of optimal settings are determined and stored. In operation, at run time, based on observation of the environment and requirements, the parameter set corresponding to the closes case are applied. Indeed, a "close enough to optimal" and thus "good enough" results can then be achieved. As a next step in the advancement towards true cognitive behavior, systems will then apply learning techniques to further calibrate and improve these heuristics. A main driver of this book is the scarcity of available spectrum. First a case considering licensed spectrum where Secondary Users use the spare wireless capacity without interfering the Primary users more efficiently has been discussed in Chap. 4. Next, application of the approach for coexistence in an unlicensed ISM band was presented in Chap. 5. Also significant energy savings can be achieved through smart(er) operation. This was illustrated in Chap. 6. As a last and most far-going study the gains achieved by adding learning and calibration to the control operation is revealed for a WLAN case in Chap. 7. For future systems, the combined consideration of both spectrum and energy efficiency will be vital to enable a sustainable future wireless Internet.

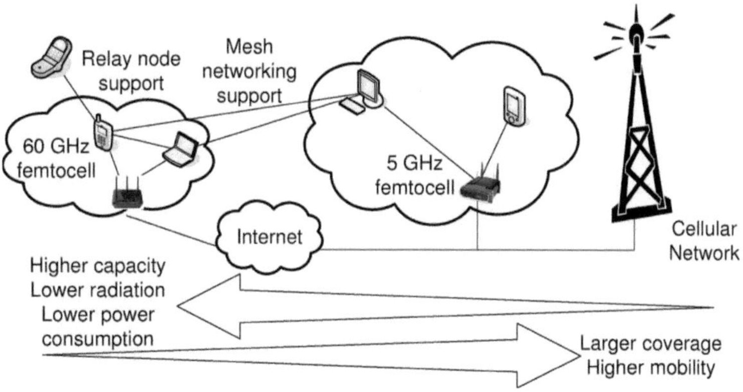

Fig. 8.2 Future network architectures go distributed for sustainable growth

8.2 Closing Remarks: The End Is Not There nor in Sight

8.2.1 Keep Moving with the Target

We closed a previous book with the statement "This is not the end, its just the beginning" [1]. Following straight engineering logics, the ideas presented in this book building on SDRs and DSA are consequently not anymore the beginning of Software Defined Radio solutions. Yet, this is definitely not the end, nor is the end in sight. In this book we advocate smart(er) and cognitive radios as an answer to the bottleneck that will occur following the ever increasing wireless data traffic within limitations of available spectrum. Recently, the energy consumption of (wireless) communication systems and the impact of the emitted electromagnetic radiation have become a cause for concern. Worldwide, initiatives have been launched and are gaining momentum to radically reduce the overall energy consumption of communication networks and to reduce the carbon footprint of the growing network [130]. If we want to build tomorrows wireless communications systems, we need to minimize the energy consumption and restrict the radiation, but not at the penalty of throughput and intelligent operation. An answer to this challenge showing great potential savings is the approach to restrict wireless communication to short ranges whenever possible. Indeed today many connections, specifically mobile phone calls, connect to a relatively faraway base-station, even if much closer access points (e.g. indoor WLAN) are present. On the longer term, one may even enable 60 GHz-based local wireless access, allowing very directive transmission at data rates that exceed 1 Gbps. This concept based on distributed access in a heterogeneous network environment is illustrated in Fig. 8.2.

A major part of the radiation and the wasted energy can be avoided and high data throughput is enabled. In order to enable this, multi-standard terminals, which can be realized attractively (both for cost and form factor) when based on reconfigurable radio solutions, will be key. Energy efficient operation of these radios is a main design goal, and cognitive features are clearly desired. The usage of mm-wave communication for short-range access purposes has recently gained interest in standardization [20]. Definitely the 60 GHz spectrum enables wider spectrum still, moreover frequency reuse at very short distances is possible, and thus smaller cells are promoted. Low cost and low power 60 GHz radios will be a vital technology, and are gaining maturity. Directive and adaptive antenna (arrays) both below 10 GHz and for mm wave will be essential to maximize connectivity for minimal energy and radiation. We believe that smart(er) and cognitive radio operation will be key to enable ever more complex wireless communication networks. We hope to have been able to show to our readers that workable solutions can bring significant gain already today. Definitely this is not the end, nor is the end in sight. As (coaches of) enthusiastic researchers we are convinced that great innovation is on its way and that this will enable more wireless capacity for/in less spectrum and energy for many years to come!

References

1. L. Van der Perre, J. Craninckx, A. Dejonghe, *Green Software Defined Radios* (Springer, Dordrecht, 2009). ISBN 978-1-4020-8210-8
2. ARCchart, Software Defined Radios in Mobile Phones (2007), http://www.arcchart.com/reports/sdr.asp
3. L. Wilson, International Technology Roadmap for Semiconductors (2009), http://www.itrs.net
4. A.L. Sangiovanni-Vincentelli, G. Martin, Platform-based design and software design methodology for embedded systems. IEEE Des. Test Comput. **18**(6), 23–33 (2001)
5. A. Pinto, A. Bonivento, A.L. Sangiovanni-Vincentelli, R. Passerone, M. Sgroi, System level design paradigms: Platform-based design and communication synthesis. ACM Trans. Des. Autom. Electron. Syst. **11**(3), 537–563 (2006). doi:10.1145/1142980.1142982
6. The Software Defined Radio Forum, http://www.sdrforum.org
7. J.G. Consultancy, SDR Market Study, SDRF-05-A-0005-V00.0, SDRForum (2005)
8. M.A. McHenry, NSF Spectrum Occupancy Measurements, Project Summary (2005), http://www.sharedspectrum.com
9. J. Craninckx et al., A fully reconfigurable software-defined radio transceiver in 0.13um CMOS, in *Proc. of the International Solid-State Circuits Conference (ISSCC)*, 2007
10. Federal Communications Commission, Policy Statement in the Matter of Principles for Reallocation of Spectrum to Encourage the Development of Telecommunications Technologies for the New Millennium (1999), http://hraunfoss.fcc.gov/edocs_public/attachmatch/FCC-99-354A1.pdf
11. Federal Communications Commission Spectrum Policy Task Force, Report of the Unlicensed Devices and Experimental Licenses Working Group (2009)
12. RSPG, Opinion on Wireless Access Policy for Electronic Communications Services (WAPECS) (2005)
13. Belgisch Instituut voor Postdiensten en Telecommunicatie, Frequentieplan (2009), http://www.bipt.be/nl/217/ShowContent/1057/Tabel/Plan.aspx
14. D. Willkomm, S. Machiraju, J. Bolot, A. Wolisz, Primary users in cellular networks: A large-scale measurement study, in *Proc. of the IEEE Symposia on New Frontiers in Dynamic Spectrum Access Networks (DySpan)*, 2008, pp. 1–11. doi:10.1109/DYSPAN.2008.48
15. J. Mitola, G.Q. Maguire, Cognitive radio: Making software radios more personal. IEEE Personal Commun. Mag. **6**(4), 13–18 (1999). doi:10.1109/98.788210
16. J. Mitola, Cognitive Radio: An Integrated Architecture for Software-Defined Radio, PhD thesis, KTH Stockholm, 2000
17. S. Haykin, Cognitive radio: Brain-empowered wireless communications. IEEE J. Sel. Areas Commun. **23**(2), 201–220 (2005)
18. S. Pollin, R. Mangharam, B. Bougard, L. Van der Perre, I. Moerman, R. Rajkumar, F. Catthoor, MEERA: Cross-layer methodology for energy efficient resource allocation

S. Pollin et al., *Software Defined Radios*, Signals and Communication Technology,
DOI 10.1007/978-94-007-1278-2,

in wireless networks. IEEE Trans. Wirel. Commun. **7**(1), 98–109 (2008). doi:10.1109/TWC.2008.05356

19. S. Pollin, Cross-Layer Resource Allocation for Quality of Service and Energy Optimization in Wireless Networks, PhD in Electrotechnical Engineering, Katholieke Universiteit Leuven, 2006. ISBN 90-5682-739-1
20. TG 802.11, Status of Project 802.11ad, http://www.ieee802.org/11/Reports/tgad_update.htm
21. B. Bougard, S. Pollin, G. Lenoir, W. Eberle, L. Van der Perre, F. Catthoor, W. Dehaene, Energy-scalability enhancement of wireless local area network transceivers, in *2004 IEEE 5th Workshop on Signal Processing Advances in Wireless Communications*, 2004, pp. 449–453. doi:10.1109/SPAWC.2004.1439283
22. D. Astely, E. Dahlman, A. Furuskar, Y. Jading, M. Lindstrom, S. Parkvall, LTE: The evolution of mobile broadband. IEEE Commun. Mag. (April 2009)
23. 3GPP TR 25.814, Physical Layer Aspects for Evolved Universal Terrestrial Radio Access (UTRA), V7.1.0 (3GPP, 2006)
24. 3GPP TR 36.913, Requirements for Further Advancements for E-UTRA, V8.0.0 (3GPP, 2008)
25. 3GPP TR 36.101, User Equipment (UE) Radio Transmission and Reception, V8.5.1 (3GPP, 2009)
26. 3GPP TS 25.913, Requirements for Evolved UTRA (E-UTRA) and Evolved UTRAN (E-UTRAN), V8.0.0 (3GPP, 2008)
27. M. Al-ayyoub, M.M. Buddhikot, H. Gupta, Self-regulating spectrum management: A case of fractional frequency reuse patterns in LTE networks, in *Proc. of the IEEE Symposia on New Frontiers in Dynamic Spectrum Access Networks (DySpan)*, 2010
28. TG 802.11, IEEE Standard for Information Technology – Telecommunications and Information Exchange Between Systems – LAN and MAN – Specific Requirements – Part 11: Wireless LAN MAC and PHY Specifications – Amendment 4: Further Higher-Speed Physical Layer Extension in the 2.4 GHz Band (IEEE-Std, 2003)
29. S.C. Ergen, ZigBee/IEEE 802.15.4 Summary, www.eecs.berkeley.edu/csinem/academic/publications/zigbee.pdf
30. Chipcon, Single-Chip 2.4 GHz IEEE 802.15.4 Compliant and ZigBee-Ready RF Transceiver (2009), http://focus.ti.com/docs/prod/folders/print/cc2420.html
31. TG 802.11h, IEEE Standard for Information Technology – Telecommunications and Information Exchange Between Systems – Local and Metropolitan Area Networks – Specific Requirements Part II: Wireless LAN Medium Access Control (MAC) and Physical Layer (PHY) Specifications (IEEE-Std, 2003)
32. Federal Communications Commission, Second Memorandum Opinion and Order in the Matter of Unlicensed Operation in the TV Broadcast Bands, Additional Spectrum for Unlicensed Devices Below 900 MHz and in the 3 GHz Band, Document 10-174 (2010)
33. ECC, Technical and Operational Requirements for the Possible Operation of Cognitive Radio Systems in the 'White Spaces' of the Frequency Band 470–790 MHz, Draft ECC Report 159 (2010)
34. ETSI, ETSI RSS, http://www.etsi.org/WebSite/technologies/RRS.aspx
35. S. Delaere, P. Ballon, Flexible spectrum management and the need for controlling entities for reconfigurable wireless systems, in *2nd IEEE International Symposium on New Frontiers in Dynamic Spectrum Access Networks (DySPAN 2007)*, 2007, pp. 347–362. doi:10.1109/DYSPAN.2007.53
36. M. Ingels, V. Giannini, J. Borremans, G. Mandal, B. Debaillie, P. Van Wesemael, T. Sano, T. Yamamoto, D. Hauspie, J. Van Driessche, J. Craninckx, A 5mm^2 40nm LP CMOS 0.1-to-3GHz multistandard transceiver, in *2010 IEEE International Solid-State Circuits Conference Digest of Technical Papers (ISSCC)*, 2010, pp. 458–459. doi:10.1109/ISSCC.2010.5433964
37. S. Thoen, Transmit Optimization for OFDM/SDMA-Based Wireless Local Area Networks, PhD thesis, K.U. Leuven, 2002. ISBN 90-5682-348-5
38. C. Schurgers, Energy-Aware Communication Systems, PhD in Electrical Engineering, UCLA, 2002

39. N. Khaled, Transmit and Receive Optimization for Mimo/OFDM-Based High-Throughput Wireless Local Area Networks, PhD thesis, K.U. Leuven, 2005. ISBN 90-5682-655-7
40. R. Jain, D.-M. Chiu, W. Hawe, A Quantitative Measure of Fairness and Discrimination for Resource Allocation in Shared Computer Systems, DEC Technical Report 301 (1984)
41. M. Dianati, X. Shen, S. Naik, A new fairness index for radio resource allocation in wireless networks, in *Proc. of the IEEE Wireless Communication and Networking Conference (WCNC)*, vol. 2, 2005, pp. 712–717. doi:10.1109/WCNC.2005.1424595
42. E. Weingartner, H. von Lehm, K. Wehrle, A performance comparison of recent network simulators, in *Proc. of the IEEE International Conference on Communications (ICC)*, 2009
43. WinLab, Rutgers University, New Jersey, ORBIT: Wireless Network Testbed (2009), http://www.winlab.rutgers.edu/docs/focus/ORBIT.html
44. Technology Center at IBBT, IBBT iLab.t (2009), http://ilabt.ibbt.be/
45. Berlin Open Wireless Lab, The Bowl Project (2009), http://bowl.net.t-labs.tu-berlin.de/
46. M. Timmers, S. Pollin, A. Dejonghe, A. Bahai, L. Van der Perre, F. Catthoor, Throughput modeling of large-scale 802.11 networks, in *Proc. of the Global Telecommunication Conference (Globecom)*, 2008
47. S.V. Gheorghita, M. Palkovic, J. Hamers, A. Vandecappelle, S. Mamagkakis, T. Basten, L. Eeckhout, H. Corporaal, F. Catthoor, F. Vandeputte, K. De Bosschere, System scenario based design of dynamic embedded systems. ACM Trans. Des. Automat. Electron. Syst. **14**(1), Article No. 3 (2009)
48. Dictionary.com LLC, Dictionary (2009), http://dictionary.reference.com
49. R. Sutton, A. Barto, *Reinforcement Learning: An Introduction* (MIT Press, Cambridge, 1998)
50. D. Mitra, An asynchronous distributed algorithm for power control in cellular radio systems, in *4th WINLAB Workshop*, 1993
51. R. Tandra, A. Sahai, SNR walls for signal detection. IEEE J. Sel. Top. Signal Process. **2**(1), 4–17 (2008). doi:10.1109/JSTSP.2007.914879
52. W.C.Y. Lee, Estimate of local average power of a mobile radio signal. IEEE Trans. Veh. Technol. **34**(1), 22–27 (1985). doi:10.1109/T-VT.1985.24030
53. K. Yonezawa, T. Maeyama, H. Iwai, H. Harada, Path loss measurement in 5 GHz macro cellular systems and consideration of extending existing path loss prediction methods, in *2004 IEEE Wireless Communications and Networking Conference (WCNC)*, vol. 1, 2004, pp. 279–283. doi:10.1109/WCNC.2004.1311557
54. C.K.H. Lim, Multicast tree construction and flooding in wireless ad hoc networks, in *3rd ACM International Workshop on Modeling, Analysis and Simulation of Wireless and Mobile Systems*, Boston, MA, August 20, 2000
55. M. Mauve, A. Widmer, H. Hartenstein, A survey on position-based routing in mobile ad hoc networks. IEEE Netw. **15**(6), 30–39 (2001). doi:10.1109/65.967595
56. F. Ye, A. Chen, S. Lu, L. Zhang, A scalable solution to minimum cost forwarding in large sensor networks, in *Proceedings of the Tenth International Conference on Computer Communications and Networks*, 2001
57. B.D. Ripley, *Statistical Inference for Spatial Processes* (Cambridge University Press, Cambridge, 1988)
58. J.S. Seybold, *Introduction to RF Propagation* (Wiley, New York, 2005)
59. TG 802.11g, IEEE Standard for Information Technology – Telecommunications and Information Exchange Between Systems – Local and Metropolitan Area Networks – Specific Requirements Part II: Wireless LAN Medium Access Control (MAC) and Physical Layer (PHY) Specifications (IEEE-Std, 2003)
60. S. Ergen, ZigBee/IEEE 802.15.4 Summary (2004), http://pages.cs.wisc.edu/~suman/courses/838/papers/zigbee.pdf
61. K.T. Le, Designing a ZigBee-Ready IEEE 802.15.4-Compliant Radio Transceiver (2004), http://rfdesign.com/mag/411rfdf4.pdf
62. A. Doefexi, S. Armour, L. Beng-Sin, A. Nix, D. Bull, An evaluation of the performance of IEEE 802.11a and 802.11g wireless local area networks in a corporate office environ-

ment, in *Proc. of the IEEE International Conference on Communications (ICC)*, vol. 2, 2003, pp. 1196–1200
63. S. Pollin, I. Tan, B. Hodge, C. Chun, A. Bahai, Harmful coexistence between 802.15.4 and 802.11: A measurement-based study, in *Proc. of the International Conference on Cognitive Radio Oriented Wireless Networks and Communications (Crowncom)*, vol. 1, 2008, pp. 1–6
64. S.-T. Center, Compatibility of IEEE 802.15.4 with IEEE 802.11, Bluetooth and Microwave Ovens in 2.4GHz ISM-Band, http://www.ba-loerrach.de
65. S. Pollin, M. Ergen, M. Timmers, A. Dejonghe, L. Van der Perre, F. Catthoor, I. Moerman, A. Bahai, Distributed cognitive coexistence of 802.15.4 with 802.11, in *Proc. of the International Conference on Cognitive Radio Oriented Wireless Networks and Communications (Crowncom)*, vol. 1, 2006, pp. 1–5
66. D. Tang, M. Baker, Analysis of a local-area wireless network, in *MobiCom '00: Proceedings of the 6th Annual International Conference on Mobile Computing and Networking* (ACM, New York, 2000), pp. 1–10. ISBN 1-58113-197-6. doi:10.1145/345910.345912
67. T.M. Mitchell, *Machine Learning* (McGraw-Hill, New York, 1997)
68. T. Chen, H. Zhang, M. Katz, Z. Zhou, Swarm intelligence based dynamic control channel assignment in cogmesh, in *ICC Workshop on Cognitive and Cooperative Wireless Networks (COCONET)*, 2008, pp. 123–128
69. C. Doerr, D. Grunwald, D. Sicker, Local indepent control of cognitive radio networks, in *Proc. of the International Conference on Cognitive Radio Oriented Wireless Networks and Communications (Crowncom)*, 2008
70. S. Pollin, Coexistence and dynamic sharing in cognitive radio networks, in *Cognitive Wireless Communication Networks* (2007), pp. 79–113. doi:10.1007/978-0-387-68832-9_3
71. E. Uysal-Biyikoglu, B. Prabhakar, A. El Gamal, Energy-efficient packet transmission over a wireless link. IEEE/ACM Trans. Netw. **10**, 487–499 (2002). doi:10.1109/TNET.2002.801419
72. R. Mangharam, R. Rajkumar, S. Pollin, F. Catthoor, B. Bougard, L. Van der Perre, I. Moeman, Optimal fixed and scalable energy management for wireless networks, in *INFOCOM 2005, 24th Annual Joint Conference of the IEEE Computer and Communications Societies*, Proceedings IEEE, vol. 1, 2005, pp. 114–125. doi:10.1109/INFCOM.2005.1497884
73. S. Pollin, R. Mangharam, B. Bougard, L. Van der Perre, I. Moerman, R. Rajkumar, F. Catthoor, Meera: Cross-layer methodology for energy efficient resource allocation in wireless networks. IEEE Trans. Wirel. Commun. **7**(1), 98–109 (2008). doi:10.1109/TWC.2008.05356
74. S. Pollin, B. Bougard, R. Mangharam, L. Van der Perre, F. Catthoor, R. Rajkumar, I. Moerman, Optimizing transmission and shutdown for energy-efficient packet scheduling in sensor networks, in *Proceedings of the Second European Workshop on Wireless Sensor Networks*, 2005
75. S. Pollin, B. Bougard, R. Mangharam, L. Van der Perre, F. Catthoor, R. Rajkumar, I. Moerman, Optimizing transmission and shutdown for energy-efficient packet scheduling in sensor networks, in *Proceedings of the 2nd European Workshop on Wireless Sensor Networks*, Istanbul, Turkey, Jan. 30–Feb. 3, 2005
76. P IEEE 802.11a, IEEE Standard for Information Technology – Telecommunications and Information Exchange Between Systems – Local and Metropolitan Area Networks – Specific Requirements Part II: Wireless LAN Medium Access Control (MAC) and Physical Layer (PHY) Specifications: High-Speed Physical Layer in the 5 GHz Band, Supplement to IEEE 802.11 Standard (Sept. 1999)
77. B. Bougard, A. Giulietti, V. Derudder, J.-W. Weijers, S. Dupont, L. Hollevoet, F. Catthoor, L. Van der Perre, H. De Man, R. Lauwereins, A scalable 8.7nj/bit 75.6mb/s parallel concatenated convolutional (turbo-) codec, in *2003 IEEE International Solid-State Circuits Conference, Digest of Technical Papers (ISSCC)*, 2003
78. D. Su, M. Zargari, P. Yue, S. Rabii, D. Weber, B. Kaczynski, S. Mehta, K. Singh, S. Mendis, B. Wooley, A 5 GHz CMOS transceiver for IEEE 802.11a wireless LAN, in *2002 IEEE International Solid-State Circuits Conference, Digest of Technical Papers (ISSCC)*, 2002

79. H. Ochiai, H. Imai, Performance of the deliberate clipping with adaptive symbol selection for strictly band-limited OFDM systems. IEEE J. Sel. Areas Commun. **18**(11), 2270–2277 (2000). doi:10.1109/49.895032
80. T.H. Lee, *The Design of CMOS Radio-Frequency Integrated Circuits* (Cambridge University Press, Cambridge, 2003). ISBN: 0-521-83539-9
81. Microsemi LX5506 InGaP HBT 4.5 6 GHz Power Amplifier
82. F.H.P. Fitzek, M. Reisslein, MPEG-4 and H.263 video traces for network performance evaluation. IEEE Netw. **15**(6), 40–54 (2001). MPEG-4 traces, http://trace.eas.asu.edu/TRACE/trace.html
83. IEEE P802.11 Wireless LANs, TGn Channel Models, IEEE 802.11-03/940r4 (May 10, 2004)
84. J. Medbo, P. Schramm, Channel models for HIPERLAN/2 in different indoor scenarios, ETSI BRAN 3ERI085B
85. R. Kravets, P. Krishnan, Application-driven power management for mobile communication. Wirel. Netw. **6**, 263–277 (2000). doi:10.1023/A:1019149900672
86. R. Mangharam, M. Demirhan, R. Rajkumar, D. Raychaudhuri, Size matters: Size-based scheduling for MPEG-4 over wireless channels, in *SPIE & ACM Proceedings in Multimedia Computing and Networking*, vol. 3020 (2004), pp. 110–122
87. A.L. Peressini, R.E. Sullivan, J.J. Uhl Jr., *Convex Programming and the Karish-Kuhn-Tucker Conditions* (Springer, Berlin, 1980), Chap. 5
88. IEEE 802.11 WG, Draft Supplement to Part II: Wireless Medium Access Control (MAC) and Physical Layer (PHY) Specifications: Medium Access Control (MAC) Enhancements for Quality of Service (QoS), IEEE 802.11e/Draft 13.0 (Jan. 2005)
89. P. Yang, Pareto-Optimization Based Run-Time Task Scheduling for Embedded Systems, PhD thesis, K.U. Leuven, 2004. ISBN 90-5682-541-0
90. M. Lacage, M.H. Manshaei, T. Turletti, IEEE 802.11 rate adaptation: A practical approach, in *Proceedings of the 7th ACM International Symposium on Modeling, Analysis and Simulation of Wireless and Mobile Systems, MSWiM '04* (ACM, New York, 2004), pp. 126–134. ISBN 1-58113-953-5. doi:10.1145/1023663.1023687
91. ns-2 Network Simulator, http://www.isi.edu/nsnam/ns
92. C. Hua, R. Zheng, Starvation modeling and identification in dense 802.11 wireless community networks, in *Proc. of the IEEE International Conference on Computer Communications (Infocom)*, 2008, pp. 1022–1030. doi:10.1109/INFOCOM.2008.156
93. M. Krunz, A. Muqattash, S.J. Lee, Transmission power control in wireless ad hoc networks: Challenges, solutions and open issues. IEEE Netw. **18**(5), 8–14 (2004)
94. I. Broustis, J. Eriksson, S.V. Krishnamurthy, M. Faloutsos, Implications of power control in wireless networks: A quantitative study, in *Passive and Active Network Measurement*. Springer Lecture Notes in Computer Science, vol. 4427 (2007), pp. 83–93
95. Y. Zhou, S. Nettles, Balancing the hidden and exposed node problems with power control in CSMA/CA-based wireless networks, in *The IEEE Wireless Communications and Networking Conference*, vol. 2, 2005, pp. 683–688
96. J.A. Fuemmeler, N.H. Vaidya, V.V. Veeravalli, Selecting transmit powers and carrier sense thresholds in CSMA protocols for wireless ad hoc networks, in *WICON '06: Proceedings of the 2nd Annual International Workshop on Wireless Internet* (ACM, New York, 2006), p. 15. ISBN 1-59593-510-X. doi:10.1145/1234161.1234176
97. V.P. Mhatre, K. Papagiannaki, F. Baccelli, Interference mitigation through power control in high density 802.11 WLANs, in *Proc. of the IEEE International Conference on Computer Communications (Infocom)*, 2007, pp. 535–543. doi:10.1109/INFCOM.2007.69
98. X. Yang, N. Vaidya, A spatial backoff algorithm using the joint control of carrier sense threshold and transmission rate, in *Proc. of the IEEE Annual IEEE Communications Society Conference on Sensor, Mesh and Ad Hoc Communications and Networks (SECON)*, 2007, pp. 501–511
99. A. Kamerman, L. Monteban, WaveLAN®-II: A high-performance wireless LAN for the unlicensed band. Bell Labs Tech. J. **2**(3), 118–133 (1997)

100. M. Lacage, M. Manshaei, T. Turletti, IEEE 802.11 rate adaptation: A practical approach, in *Proceedings of the 7th ACM International Symposium on Modeling, Analysis and Simulation of Wireless and Mobile Systems*, 2004
101. G. Holland, N. Vaidya, P. Bahl, A rate-adaptive MAC protocol for multi-hop wireless networks, in *Proceedings of the 7th Annual International Conference on Mobile Computing and Networking*, 2001
102. B. Sadeghi, V. Kanodia, A. Sabharwal, E. Knightly, Opportunistic media access for multirate ad hoc networks, in *Proceedings of ACM MobiCom '02*, 2002
103. B. Chakravarty, Rate Control Algorithms for IEEE 802.11 Wireless Networks, Master's thesis, The University of Texas at Arlington, 2007
104. J. Won, C. Kim, A downlink rate adaptation scheme in IEEE 802.11 WLANs using overhearing, in *Intl. Conf. on Information Networking ICOIN '08*, Busan, Korea, 23–25 Jan. 2008, pp. 1–5
105. J. Choi, J. Na, Y. Lim, K. Park, C. Kim, Collision-aware design of rate adaptation for multi-rate 802.11 WLANs. IEEE J. Sel. Areas Commun. **26**(8), 1366–1375 (2008)
106. S. Khan, S. Mahmud, K. Loo, H. Alraweshidy, A cross layer rate adaptation solution for IEEE 802.11 networks. Comput. Commun. **31**(8), 1638–1652 (2008)
107. S. Lee, K. Chung, Combining the rate adaptation and quality adaptation schemes for wireless video streaming. J. Vis. Commun. Image Represent. **19**(8), 508–519 (2008)
108. H. Jung, K. Cho, Y. Seok, T. Kwon, Y. Choi, RARA: Rate adaptation using rate-adaptive acknowledgment for IEEE 802.11 WLANs, in *Proceedings of the Consumer Communications and Networking Conference*, 2008
109. J. Kim, S. Kim, S. Choi, D. Qiao, CARA: Collision-aware rate adaptation for IEEE 802.11 WLANs, in *Proc. of the IEEE International Conference on Computer Communications (Infocom)*, 2006, pp. 1–11
110. S. Wang, A. Helmy, BEWARE: Background traffic-aware rate adaptation for IEEE 802.11, in *Proceedings of the 9th IEEE International Symposium on a World of Wireless, Mobile and Multimedia Networks (WoWMoM)*, Newport Beach, CA, June 2008
111. Y. Yang, J. Hou, L.-C. Kung, Modeling the effect of transmit power and physical carrier sense in multi-hop wireless networks, in *Proc. of the IEEE International Conference on Computer Communications (Infocom)*, 2007, pp. 2331–2335. doi:10.1109/INFCOM.2007.275
112. Z. Zeng, Y. Yang, J. Hou, How physical carrier sense affects system throughput in IEEE 802.11 wireless networks, in *Proc. of the IEEE International Conference on Computer Communications (Infocom)*, 2008, pp. 1445–1453
113. F. Rossetto, M. Zorzi, Enhancing spatial reuse in ad hoc networks by carrier sense adaptation, in *Proc. of the IEEE Military Communications Conference (MILCOM)*, 2007, pp. 1–6. doi:10.1109/MILCOM.2007.4455087
114. E.B. Koh, C.-K. Kim, Mitigating starvation in CSMA-based wireless ad hoc networks using carrier sense threshold, in *Proc. of the International Conference on Software, Telecommunications and Computer Networks*, 2007, pp. 1–6
115. H. Ma, S. Shin, S. Roy, Optimizing throughput with carrier sensing adaptation for IEEE 802.11 mesh networks based on loss differentiation, in *Proc. of the IEEE International Conference on Communications (ICC)*, 2007
116. X.-H. Lin, Y.-K. Kwok, V. Lau, Power control for IEEE 802.11 ad hoc networks: Issues and a new algorithm, in *Proc. of the International Conference on Parallel Processing*, 2003, pp. 249–256
117. B. Alawieh, Y. Zhang, C. Assi, A distributed power and rate control scheme for mobile ad hoc networks, in *Proc. of the International Symposium on Modeling and Optimization in Mobile, Ad Hoc, and Wireless Networks*, 2008, pp. 335–343. doi:10.1109/WIOPT.2008.4586087
118. T. Kim, H. Lim, J. Hou, Improving spatial reuse through tuning transmit power, carrier sense threshold, and data rate in multihop wireless networks, in *Proceedings of the 12th Annual International Conference on Mobile Computing and Networking*, 2006
119. H.P. Young, *Strategic Learning and Its Limits* (Oxford University Press, Oxford, 2005)
120. Y. Shoham, R. Powers, T. Grenager, Multi-agent reinforcement learning: A critical survey, in *AAAI Fall Symposium on Artificial Multi-Agent Learning*, 2004

121. L. Busoniu, R. Babuska, B.D. Schutter, Multi-agent reinforcement learning: A survey, in *Proc. of the 9th International Conference on Control, Automation, Robotics and Vision*, 2006, pp. 1–6. doi:10.1109/ICARCV.2006.345353
122. L. Busoniu, R. Babuska, B.D. Schutter, A comprehensive survey of multiagent reinforcement learning. IEEE Trans. Syst. Man Cybern., Part C, Appl. Rev. **38**(2), 156–172 (2008)
123. S. Mannor, J. Shamma, Multi-agent learning for engineers. Artif. Intell. **171**(7), 417–422 (2007)
124. S.O. Kimbrough, M. Lu, Simple reinforcement learning agents: Pareto beats Nash in an algorithmic game theory study. ISeB **3**(1), 1–19 (2005)
125. S. Hart, Adaptive heuristics. Econometrica **73**(5), 1401–1430 (2005)
126. Y. Park, Y. Seok, N. Choi, Y. Choi, J. Bonnin, Rate-adaptive multimedia multicasting over IEEE 802.11 wireless LANs, in *Proc. of the Consumer Communications and Networking Conference*, vol. 1, 2005, pp. 178–182
127. M. Timmers, S. Pollin, A. Dejonghe, L. Van der Perre, F. Catthoor, Exploring versus exploiting: Enhanced distributed cognitive coexistence between IEEE 802.11 and IEEE 802.15.4, in *Proc. of the IEEE Conference on Sensors*, 2008
128. G. Bianchi, Performance analysis of the IEEE 802.11 distributed coordination function. IEEE J. Sel. Areas Commun. **18**(3), 535–547 (2000)
129. K. Fall, K. Varadhan, VINT project, The ns Manual, UC Berkeley, LBL, UCS/ISI, and Xerox PARC, http://www.isi.edu/nsnam/ns/ns-documentation.html
130. www.greentouch.org

MIX
Papier aus verantwortungsvollen Quellen
Paper from responsible sources
FSC® C105338

If you have any concerns about our products,
you can contact us on
ProductSafety@springernature.com

In case Publisher is established outside the EU,
the EU authorized representative is:
Springer Nature Customer Service Center GmbH
Europaplatz 3, 69115 Heidelberg, Germany

Printed by Libri Plureos GmbH
in Hamburg, Germany